Time, Light
and the
Dice of
Creation

Praise for
Time, Light and the Dice of Creation

'Philip Franses is a pilgrim in life and thought. His book is an invitation to explore the cracks between the hard facts of science. It looks for meaning rather than measurement. We realise that the journey itself is the goal. The message of this book is a way of living.'

Shantena Augusto Sabbadini, physicist, teacher and writer,
Associate Director of The Pari Center of New Learning, Italy

'Franses' rich and insightful exploration of the tensions in scientific inquiry between wholeness and its constituent parts mirrors beautifully the two ways of understanding reality in deep Buddhist meditation. Absolute Reality – oneness or wholeness – is experienced by investigating appearances which appear separate. What feels like a tension or paradox between the particular and the Absolute is actually an experience of the same thing, viewed from different perspectives. There is no paradox because there is no opposition – everything manifests in its separateness and in its wholeness simultaneously.'

Margaret Wheatley, author of Leadership and the New Science
and many other books

Time, Light
and the
Dice of
Creation

Through Paradox in Physics
to a New Order

Philip Franses

Floris Books

First published in 2015 by Floris Books
© 2015 Philip Franses

Philip Franses has asserted his right under the
Copyright, Designs and Patent Act 1988 to be
identified as the Author of this work

 This book is also available
as an eBook

British Library CIP Data available
ISBN 978-178250-172-5
Printed in Poland

Contents

Introduction   7

Acknowledgments   17

The Dice of Existence   19
   1. Division faces unity: quantum theory   20
   2. Darkness faces light: relativity   46
   3. Past faces future: electromagnetism   76

The Dice of Renewal   105
   4. Chance faces order: the fall   106
   5. Potential faces expression: the arising   133
   6. Emptiness faces form: the hub and the rim 160

Creation   195
   7. Energy faces time: myth   196

Bibliography   227

Index   232

Introduction

The light of wholeness

The title and cover of this book reflect an aspiration to recover the light of wholeness. Wholeness, it is said, contains everything about itself, within itself. Wholeness is a quality one recognises as the impetus to one's actions, or the motivation behind the scenes, without ever revealing itself fully. Wholeness is what one meets on a journey, as the source and end of adventure.

In the second half of the nineteenth century, science set out to discover a path, about which everything at every stage could be exactly known. The sense of mystery into which one journeyed was replaced by the static elements of reason by which one could explain the world. The twentieth century needed an instrument to harness wholeness into useable theorem. So a method was developed to tease this light of wholeness into the dark containment of a pure rationality.

The debate around quantum theory naturally focused on wholeness, since there was no complete definition of individual existence independent of context. Wholeness was conceived of as a mathematical artefact. The world only had reality as it was brought into being through the act of measurement. Measurement is a very limited application of something much wider which is 'seeing'. True seeing is not about the measurement of a quantity, but the engagement into a situation to discover its 'meaning' – that is, the convergence of content and happening on a form.

So having declared measurement as the way of interacting with the world, very conveniently one has dropped off the end of physics without anyone noticing all those difficult possibilities to do with unmeasurable routes to meaning.

It seems then that we have done the impossible: we have placed the realisation of meaning inside knowledge and nothing can ever happen to disprove this assumption! It seems a cast-iron case. Only the expectation that knowledge already has about the world is a legitimate basis for experience.

Once physics has made the step of sealing meaning into knowledge, nothing can ever happen to test whether that step was indeed justified. After all, our knowledge is our own, our institutions are free to set their own academic standards, so what else is there to even challenge whether this step was legitimate?

Surely then this book is simply another commentary on the futility of the move to seal meaning into knowledge, unable as we are to penetrate into the protective containment of our conceptual hold on the world? The attempt we make to face this paradox cannot be solved from within the method of cause and effect.

As every child knows, wholeness is the route to life. When I was six years old, I had a desk. I had many precious things in it and I thought, 'Well, it is not really safe having them all in this desk because I have a key here... And though it's locked, somebody can take the key and unlock it.' I had the idea of putting the key into the drawer that was locked, through a small crack that was there. So, nobody could get to it.

My desk locked by a single key, became securely protected at the moment when, fitting the key through the slit between the locked drawers, all opening was prohibited. This event reflects what has happened to science. Rather like my desk drawer with key inside it, the conceptual world of science has closed about its own capability to understand precisely what its theories may statically describe.

The present book reopens the question of meaning that lives in paradox and retakes a path into the unknown, to find where we are beyond mere measurement.

Method

The chapters which follow, like the faces of a dice, are not chronological, but each opposition fits to add to the significance of the previous ones. The story is built up through the points of view of physicists that reflect one another. The understanding accumulates through reflections that play on one another, to give a suggestion of the whole quality pointed to beyond. All the perspectives develop an understanding about the world and our place in it, without having to fix what it is we are describing. The book is as a juxtaposition of faces that together hold the chance of existence acting into order.

The first section is called *The Dice of Existence*. This section details a series of opposing faces in the history of physics. The mirror reality of the atom is reflected in the many faces implicated in its discovery. Bohr meets with Heisenberg on the edge of global destruction; Pauli meets with Jung in the dreams of creation; Bohm meets with Bortoft in the quest of existential unification. The next chapter jumps to the origin of physics. Newton meets Leibniz in the courtroom of the Royal Society to judge over who first invented calculus. The dispute of philosophies is between Newton's insight into material motion and Leibniz's foresight of the existential fulfilling of unity. Einstein resolves the dispute by placing another mirror on the world, the immutable speed of light. The third chapter is about Maxwell's theory of electromagnetism. Light has two faces, one seeding a causal and the other a prophetic account. The lived tension of these oppositions is illuminated throughout in the way significance writes chance into coherence.

In a dice, the separate faces require opposition to present their particular possibility. And similarly here, the paradoxical opposition of various theories and predictions of physics face their partial perspectives to a deeper, whole truth of experience.

In the second section, *The Dice of Renewal*, the future is engaged as a practical test. The action moves to fateful happening concerned directly with the world. Right at the heart of physics, chance is the yeast that brings through the established parts of the first section, the movement into the whole of the second section. The method knows from within its own process how the partial elements of potential are drawn into whole expression. The logicians Spencer Brown and Lou Kauffman, in the sixth chapter, see the world as a movement whose arbitration is either wholeness or emptiness. Kauffman describes this as 'the concept of a system whose structure is maintained through the self-production of its own structure.' (Kauffman 2015, p. 12). The way of working with this logic is to travel inward through the method's paradoxical premise, making connection until arriving at the essence of nature that retrospectively defines where we have come from. This method whittles away towards an essence that is the worth of the journey of inquiry.

In the third section, *Creation*, time faces energy in the specific account of meaning. The last chapter finds an unexpected third reflection on the story of the first chapter. Carl Friedrich von Weizsäcker travelled with Heisenberg to meet Bohr in Copenhagen to discuss the feasibility of developing an atomic bomb. Von Weizsäcker later went on to write *The Unity of Nature*. In this book, he exposes the shortcomings of a normal view of time in physics. Instead he suggests returning to an alternative conception of time that fulfils the separate faces we have explored as a new holder of events, not only between the concepts and ideas of physics, but between physics and theology!

The Dice of Existence

We travel through paradoxical dimensions. The understanding through this book is that where there are tensions in physics, as in descriptions through unity or division, darkness or light, past or future, and so on, physics finds a way to balance a theory at the very edge of this tension, where both states are ambiguously possible.

In Chapter 1, Bortoft invokes phenomenology by entering the paradox of seeing both division and unity as simultaneous attributes of quantum phenomena.

←——————————— division/unity ———————————→

In Chapter 2, Einstein takes the tension between Newton's view of matter as dominating over change and Leibniz's view of change realising itself in unities of dynamic identity. Einstein plays within the paradox of having matter itself operate on the medium of space and time through which matter moves. The result is relativity.

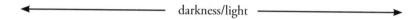

darkness/light

In Chapter 3, Maxwell's equations have two solutions: the retarded forward-moving wave of light and another advanced backward resolution to a formative darkness. Feynman and Wheeler's solution is to combine these tensions in the balance of a dynamic illumination. Their model is taken up in the quantum potential field and the bi-metric theory of relativity, as explanation for dark matter.

past/future

Physics, in each case, works with paradox by applying at one stage of an argument, one direction of the ambiguous alternative, and at the other stage, another. So in quantum theory, the aspect of unity where all potentials are considered universally overlaps with the statement of each individual separately. In relativity, light is both universal and specific. For Feynman and Wheeler, radiation goes both forward from the past and backward to the future.

Each of these paradoxes relate to the others. Although they are applied individually in terms of quantum theory, relativity and electromagnetism, their dimensions of tension also work in combination.

There is a relatedness of form between the ambiguous tensions that contains the essence of each paradoxical statement in the other two. The paradox of division/unity does not complete itself in its own statement of quantum theory. Rather the paradoxical nature of this one ambiguity requires the other two ambiguities to fully frame its own power of application. The world does not collapse through theories into a self-explanation. The experience of the spirit is to fulfil all these independent articulations in a composite story of unity. The relatedness is not in a common factor to which the three theories reduce; the relatedness is the quality of an association that expands experience from fragmentation into unity (see *Figure 1*).

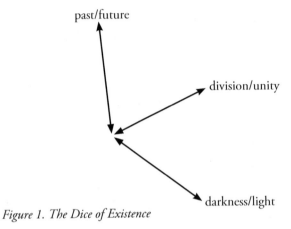

Figure 1. The Dice of Existence

The Dice of Renewal

We can now do something similar with the second section. Each of these chapters/dimensions is a similar paradoxical element that proposes an ambiguous route of becoming.

The nature of the dice combining paradoxical tenets is that it adds chance into an order that is more than just a linear deduction of earlier states, as shown in Chapter 4.

The fulfilment of chance brings whole form to manifest differently according to circumstance, as shown in Chapter 5.

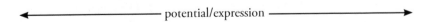

Right at the heart of physics, chance is the yeast that brings through the established parts of the first section, the movement into the whole of the second section. For Bohr, probability or chance was central to the particle world. The section, *The Dice of Renewal*, explores chance as the risk of the individual to realise through its journey the outcome of wholeness. The individual surrenders itself, outside any pre-defined structure, to a journey of discovery, where order is introduced only at the end by the fulfilment of wholeness. Chance ceases to be something mathematical, that which contains existence in the predictive walls of a theory, and becomes something experiential, that which transforms the finite through surrender to an infinite outcome. Bortoft, Goethe, Hiley, Kauffman, Spencer Brown, all developed science to navigate this other notion of chance. The surrender from initial definition delivers the partial elements of potential into whole expression. As cited above, Kauffman describes this as 'the concept of a system whose structure is maintained through the self-production of its own structure'.

emptiness/form

Again we put these together into the Dice of Renewal (see *Figure 2*).

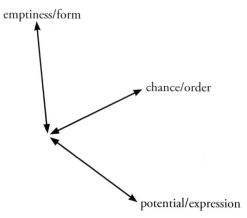

Figure 2. The Dice of Renewal

Creation

To combine these two statements of existence and renewal, we use a third step of Creation, involving the holding of the paradoxical ambiguity between energy and time (see *Figure 3*).

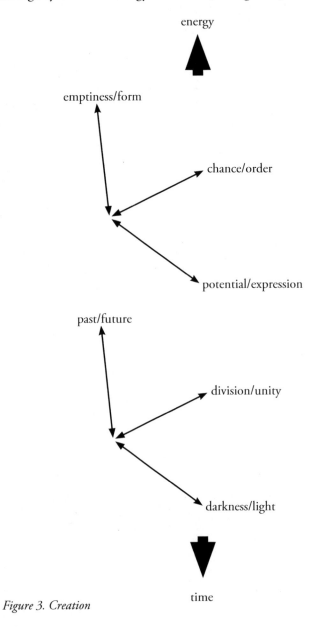

Figure 3. Creation

Creation makes the double movement between mystery and causality. Two aspects of time, the mythological and the chronological, oversee the whole cycle of existence, in the fulfilling of the experiential, participative journey into the nature of wholeness.

Journey

One day in a bookshop, in 1994, I happen to stray from the physics section into the neighbouring theology section. Somehow as happens in these cases, the whole feeling of a journey to be undertaken is communicated to me in one feeling of connection. All the turbulence between the questions of the physics and theology sections come to rest in this one offered resolution.

The instances of my experience to do with science (learning, puzzling, solving) and those to do with spirit (travelling, discovery, rebirth) live a movement where something becomes apparent before the separation into the two disciplines.

My first impression is concentrated, complete, a vision of something full and finished. The feeling lacks any structure. It is the whole, in its full exhibition, without any division into parts. The concepts which I need to articulate the feeling are totally missing. The first impression impels me to work towards the realisation. My involvement is actively included as participatory factor. The unity of what world and writer can achieve together discloses a genuine and novel illumination out of darkness.

After this introduction, I begin to visit the bookshop regularly. Gradually the form of the book begins to emerge through specific questions that become in their turn the focus of more broad explorations. The parts form themselves through those questions that carry the whole meaning, as disclosed in the first impression. The concepts that are to give structure to the project become apparent.

For instance, one night in 1995 I cannot sleep trying to understand where the split between science and spirit happened. My thought focuses on Bohr as the critical character in this separation. The next day, with little rest, I travel to the bookshop. The first series of books

I meet on the shelf is *The Philosophical Writings of Niels Bohr* in three volumes. It is brought home to me, without being able to articulate fully at the time, that Bohr mixes up two questions: 'what is there?' and 'what happens?' The definition determines henceforth the scope of what science sets itself up to explore.

The parts of the book are presented as questions, or enigmas, that carry in them the essence of the whole project. The parts are ideally suited to deliver the full meaning, but their characters are also derived from the whole. Each of the three main theories of physics – quantum theory, relativity and electromagnetism – holds in itself the question of the interpretation of physics.

All the chapters that you are going to read are not trying to build up a single argument about physics. They have each suggested themselves as dynamic ways to carry the question of the interpretation of physics in its entirety to collective significance.

My encounter in the bookshop initiated me on a journey through physics, using my mathematics degree as an aid to navigate me on what was at the start, completely unchartered terrain. It felt like one of those moments where one sets out into the night, realising it is very unlikely one will get a lift, but being drawn nevertheless by the adventure into the challenge of the forbidding journey ahead.

Acknowledgments

I am especially grateful to: Colin Ashton; Kate Ashton; Emilios Bouratinos; Gianlauro Casoli; Valerie Charlton; Chris Clarke; Satish Kumar; Alice Oswald; Peter Oswald; Shantena Sabbadini; Harvey Schoolman; Patricia Shaw; Rupert Sheldrake, who all provided key advice, assistance and perspective at various stages of the writing.

A special thanks to Minni Jain, who provided continual belief in the project and gave invaluable insight at critical stages.

In 2005 I had completed the one year residential MSc in Holistic Science at Schumacher College, Devon, in the UK. I would like to thank all teachers, students, staff and volunteers at Schumacher College who shared this journey.

An influential teacher and later colleague and friend, was Henri Bortoft. Henri Bortoft first introduced me to the reality of the living relation of whole and part. His dedication to describe and communicate existence as a dynamic appearing are articulated in his book *Taking Appearance Seriously* (Bortoft 2012) published two months before his passing away. I owe deep gratitude to Henri and Henri's wife, Jackie Bortoft, who read through this manuscript on a number of occasions.

My dissertation, supervised by Brian Goodwin, was called *Living Ambiguity, Contextual Choice in the Outcome of DNA* (Franses 2006). Brian Goodwin was a writer and academic. After Brian's passing in 2009, I became faculty lecturer on the MSc in Holistic Science alongside Stephan Harding. I would especially like to thank Brian and Stephan, for their collegiality and wisdom.

It was through writing a chapter in *The Intuitive Way of Knowing: A Tribute to Brian Goodwin*, (Floris Books 2013) that I came into contact with the editor Christopher Moore. Apart from his reading endless versions of the text and pointing the way forward, a particularly poignant moment was when preparing this book for publication. In a moment of turmoil, where ideas seemed scattered in disarray, Christopher remained completely calm, listening to everyone. Suddenly order took hold and the book had arrived at its form in the world. Christopher Moore, thank you!

The Dice of Existence

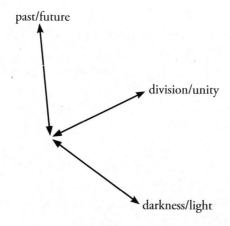

Chapter 1

Division faces unity: quantum theory

One of the people working on the atom in the early days around the time of the First World War, was Niels Bohr, a theoretical physicist. After making important early contributions when working with Rutherford in Manchester in 1912, Niels Bohr continued to seek a theoretical basis to the atom. However from 1915 to 1918 he published nothing:

> I know that you understand ... how my life from the scientific point of view passes off in periods of over-happiness and despair, of feeling vigorous and overworked, of starting papers and not getting them published, because all the time I am gradually changing my views about this terrible riddle which the quantum theory is. (Bohr 1918)

Reading his letters you get this feeling of a desperate soul, because when trying to understand the atom he's trying to understand existence at its furthest shore of where it comes into being. Bohr was completely perplexed with this effort, partly because he had been schooled in Newtonian physics, he thought everything should be reducible to space and particles and forces. And partly because he had no holistic training which could prepare him for the task.

> I suffer from an unfortunate inclination to make results appear in systematic order. (Bohr 1919)

He had no way to understand this realm which did absolutely nothing – it only appeared when you did something to it. It was completely mute because it was the beginning of existence, the elementary shore of existence. Nothing actually happened to it until you did something to it, and then it would respond and it would say 'Yes, I'm the shore of existence.' But if you just left it there it would do absolutely nothing because it would then say, 'I'm the shore of existence, if you don't bring me into existence I'm not here.'

Bohr was going round and round with these ideas and in his letters you get an idea that he was wrestling with this darkness that wouldn't let him in, it wouldn't let him understand, it wouldn't let him explain. And then he had the idea of not trying to understand it in a Newtonian way, that this happened, and this happened, and this happened; but trying to say what he did understand and what he didn't.

Concepts

The whole western worldview is based on this idea that inside the human mind there are these concepts like mass, space, time, speed. These concepts are intellectual constructs. The aim is that through these intellectual constructs man will be able to substitute this constructed conceptual world for reality. And the advantage of the concept is that it is man-made, so once you have really understood how that concept acts in reality, it allows you to control how you want to work with it.

So once you've understood the concept of Newton's laws of motion, you can build a steam engine, or a locomotive or a train for the concept allows you to catch hold of how reality works and then, instead of encountering an unknown reality, it allows you to meet something you know, you can predict, you can control and you can understand. And similarly once you understand the concept of electromagnetism, then there can be millions of people using mobile phones (which is quite incredible if you just imagine, two hundred years ago, how it would have seemed: everybody walking around talking and listening to people the other side of the world).

The goal of the Enlightenment era of science was that if we pursued this conceptual approach, we would eventually get to the super-conceptual framework from which we could construct a totally useable world, a world we totally understood and could totally represent within these thought realities. The concept in our mind, substituted for reality, allowed us to be forever the master of that reality. We could train people from a young age in substituting reality for this conceptual thought world; we could improve technology more and more; and gradually we could create this far superior world where instead of there being disease, premature death and suffering, we would have a direct understanding through our concepts of how the world worked and so we could master reality in all its different aspects.

It was long thought, even from Greek time, that if you understood the world deeply enough you'd eventually get to the concept of the atom, the lowest common denominator. Just as in knowledge you'd finally get to a concept that would explain everything else, so you could reduce the world and finally get to this single thing that underpinned everything else. Everything was made of atoms. So if we go back and back and back, and divide, divide, divide, we'd eventually get to these tiny things and these would be the building bricks of the world.

Bohr's progress

Two contradictory aims were demanded from the study of the atom in 1900. The atom was on the one hand a Utopian goal of the Enlightenment project. The belief was that rationality would be able to establish a self-consistent theory of everything starting from elementary building blocks. On the other hand, the atom was an overarching conceptual term for a host of confusing experimental data about a mostly empty space with occasional orbiting electrons and a central heavy nucleus made of protons and neutrons.

The atom was first envisioned a bit like the solar system, with a positive nucleus and the negative electrons orbiting around it. You knew there was something in the middle, very small, and there were these orbits around it of electrons. It seemed like a very good model.

But there was a problem with this model. If a negatively charged electron moves around the positively charged nucleus, it should be giving off radiation and gradually losing energy and falling into the centre. So why did the atom remain this very strong thing at the basis of all matter?

There was another mystery, that each element had a particular frequency of colour that was absorbed or transmitted. When you looked at the stars you could identify the elements in the stars from the frequency of the light each element absorbed.

The results were baffling from a normal scientific perspective. Bohr, in 1912, was the first scientist to come up with a model of the atom that fitted all of these enigmas together.

Bohr's 1912 paper said you could explain the frequencies of energies if you stopped focusing on the atom as something fixed and just focused on the transitions between the various energy levels which the electron could occupy. The atom doesn't really exist, that is why there is no energy being given off when nothing happens to it. But when light enters, or when something allows a transition from one energy level to another of the electrons, then a transition will occur and only by that do we know of the atom's existence. It isn't a thing but it is more a dynamic that responds when light of the right frequency comes in, to allow the jump between this outer energy level and the inner energy level. You could explain the atom not as some *thing* but more as a kind of happening that, in relation to its context, made these jumps allowed by the energetic system. No thing but a happening in response to its environment.

In Bohr's model there are only a number of stationary states in which an electron can exist. An electron when excited by energy coming into the system can make a jump from one energy state to another.

So when the system is in a stable state, it doesn't interact with the environment and lose energy as you would expect. The electron reveals its relation to the environment only when the system jumps to another energy state. And because each element has a particular set of allowable orbits of its electrons, the energy jumps it makes between different levels happen only with photons of particular frequencies – so each element is associated with particular colours of excitation or absorption. The spectrum of colours of each element correspond to the frequencies predicted by the jump in energy levels.

In the absence of light, the inner being is collapsed into a total isolation, unknowable to any outside inquiry. In this sense the electron in its stable state is hidden energetically from the world. However when a transition of the electron subtly represents a change of energy in the whole system, the light that is emitted or absorbed, is the window on the world the electron needs for its experience to exist.

These windows of jumps in the energy of the electrons to reflect the change in the whole state of the atom are also the only windows on the existence of the atom. There is absolutely nothing telling you that the atom is there. On the other hand, when the electron makes a transition between equilibrium states, then it gives a sign of its presence. So if you prod the electron in such a way that it jumps to a new state, then you can see the atom is there. You bring that nothingness into existence and suddenly it says, 'Hello! Here I am! I'm the shore of existence.'

As Arthur Zajonc writes about light:

> Sense objects must possess sharply defined attributes. Light, quantum mechanically considered, need not. Its attributes are more holistic; in general they exist in inseparable or entangled combinations, at least until the moment of measurement, whatever that is.
>
> What are the primary qualities of light that vouchsafe its unambiguous existence? The extraordinary response given by quantum realism is that there are none. Light, as an enduring, well-defined, local entity vanishes. In its place a subtle, entangled object evolves, holding all four of its quantum qualities suspended within itself, until the fatal act of measurement. (Zajonc, p. 315)

Double-slit experiment

The peculiarity of the quantum domain persisted. There were many ironies. For instance J.J. Thompson at Cambridge in 1897 first came to the discovery of the electron as a particle for which he was awarded the Nobel Prize in 1906. Meanwhile his son George Thompson, also at Cambridge, won the Nobel Prize in 1937 for his discovery of the wave properties of the electron using the technique of electron diffraction. The duality of particle/wave is illustrated in the experiment below, where a particle is shown to exhibit both its particle and its wave nature at the same time.

The apparatus is set up so that a particular photon (or other particle such as an electron) may travel through either one of two slits on the way to a screen that records its point of impact. (See *Figure 4.*)

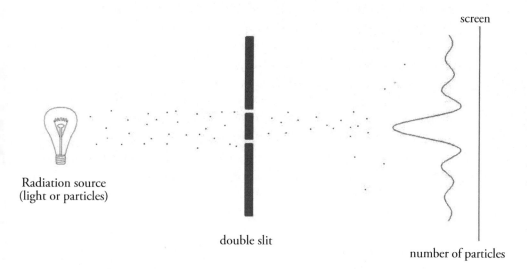

Radiation source
(light or particles)

double slit

screen

number of particles

Number of particles registering at screen appears to show information from both slits in each individual trajectory.

Figure 4. Double-slit interference experiment showing wave-particle duality

We choose to examine a phenomenon which is impossible, *absolutely* impossible, to explain in any classical way, and which has in it the heart of quantum mechanics. In reality, it contains the only mystery. (Feynman 1965a)

The apparatus reproduces an original experiment of Young's, where light passing through two slits will produce interference bands. This was used by Young to demonstrate the wave nature of light. The pattern can be explained as the interference of the troughs and peaks of the waves passing through the separate slits. On the other hand in quantum theory one has to deal with light as particle. The experiment can reduce the intensity of light until only one particle of light, a photon, is in the apparatus at one time. The incidents of light on the screen for each particle in the apparatus register at a unique place, as one would expect with a particle. But as one continues the experiment, the individual instances build up into an interference pattern as if one was dealing with a wave.

Arthur Zajonc describes a slightly modified experiment of Pfleegor and Mandel in 1967 using two lasers with a very small angle between them to fire individual photons that could be detected on a screen. Over very short time periods, interference patterns could be clearly seen. This could be explained by the wave nature of light. They then reduced the light intensity until only one photon was in the experimental apparatus. A weak interference pattern was still visible.

In the original experiment, you fire a few photons through the two slits and they make marks on the screen, so they are going through one slit or the other. Then you fire a few more photons, so they must go through one slit or the other because they're single photons. Then you see more of them hitting the screen. Then you introduce a few more photons and you finish the experiment and you find an interference pattern of light on the screen. So even though you've been doing the experiment particle by particle and the photon must have gone through one slit or the other, what emerges is a pattern of interference as if it's a wave. And even though it has been separated over time, it is as if somehow the possibility of what could happen at the other slit is affecting what happens at the slit the photon *does* pass through.

This implies that it is not possible to say through which slit the particle has gone (from which laser a single photon is produced). Indeed as Arthur Zajonc shows, 'the single quantum is being "co-produced" in the two different lasers.'

The identity of the photon is thus only present as a potential until the moment of observation. As Greenstein and Zajonc conclude:

> The metaphysical implications are profound. The experimental tests go so far as to change the very way we should think of physical existence at its most fundamental level. We must think [of the micro-world] in terms of non-locality, and/or we must renounce the very idea that individual objects possess discrete attributes. (Greenstein & Zajonc, p. 161–62)

Further about non-locality:

> Imagine that you are conducting experiments on what appears to you to be an isolated particle. But is it an isolated particle? Even though nothing is touching it, even though nothing else is even near, it might be entangled with some other particle. Furthermore, there is no local experiment you can perform to tell whether or not this is the case. (Greenstein & Zajonc, p. 169)

The action of the individual particle is dependent on the context of its passage. It is also possible to determine exactly through which slit the particle travelled before it reaches the screen. By measuring through which slit the particle passes, the interference pattern disappears. The particles in this case aggregate opposite the slits, passing cleanly through one or other slit. The photon having been revealed to its context amongst things earlier, thereafter takes on the behaviour of the particle.

The choice of where the observation is made affects how a particle manifests.

Appearance of a solution

Bohr was the figure given the responsibility to marry the cultural need to see in the atom a fundamental concept and the weirdness of the phenomena that experiments like the one above reported. He was faced with two seemingly irreconcilable pressures: the ambition to see in the atom a simple answer that could found the Enlightenment project on sound rational principles from then on, and the science that seemed to show the atom as behaving outside all previous notions of predictable law.

Bohr's genius was in coming up with a solution acceptable to both camps. Bohr concentrated not on what was happening to the photon on its journey through the slits, but on how the particle appeared to the act of measurement. He changed the character of science from being a representation of what happened in an outside reality to a study of how reality manifested whenever we asked of it a question.

In Bohr's subtle shift of attention to the act of measurement, a predictive capability of the probability of where the particle would manifest, still left science with the feeling that it was master of the atomic system. It left intact a coherent mathematical structure of prediction (at least to the extent of probabilities of values occurring within the total situation of the system).

> The procedure of measurement has an essential influence on the conditions on which the very definition of the physical quantities in question rest. (Bohr 1935, p. 1024)

Bohr realised there was huge potential in the subject for disguise, because the atom or the particle only revealed itself when you prodded it, or you disturbed it, or you measured it. There was this huge potential for not saying exactly who or what you were and you could get away with it because you only had to reveal who you were when you were measured. So you could completely hide what you were in between, as long as when someone asked you, you had an answer, 'I'm the photon, with this position,' or 'I am the electron, with this momentum.' In the meantime you could be anyone you wanted. If someone has an expectation of your identity, then you could answer according to that expectation. There is a huge possibility for disguise.

Bohr led other scientists to the Copenhagen interpretation of quantum mechanics (Bohr's institute where much of the work was done was in Copenhagen). The interpretation stood up to intellectual rigour, recognisable to classical science. There was this wave (Schrödinger wave function) of potential that existed containing preparatory information about all the different conceptual possibilities of what could be measured. When you performed a measurement, this wave collapsed and one of the possibilities would be chosen. The ingenious thing was that by focusing on the act of measurement, everything was seen in relation to the act of measuring itself. You didn't have to consider the nature of the preparatory state in which the particle was half-existing and half-not-existing.

Bohr held something firm in his hand. Whenever you made a measurement, something appeared, and you had a good clue about what appeared, because of this wave function and its probabilities. You still held on to this notion that reality adhered to a conceptual substitution capturing how it behaved. You still understood the electron as a negatively charged thing; you still understood position and momentum, as if these were a value substitute for reality.

The theory gave rise to the illusion that the conceptual world had finally understood absolutely the foundation of itself. In the attempt to understand the atom that underlay everything else, it was as if theorists had also discovered where concepts arose. The wave function produces this non-real probability, which at a particle level tells how concepts as position, momentum and mass manifest. It gave the impression that science had arrived at something final.

As Max Born said in 1926 in *The Structure of the Atom*:

Up to the present, in physics as in other sciences, every result that one age has proclaimed as absolute has had to fall after a few years, decades or centuries, because new investigations have brought new knowledge and we have become used to consider the true laws of nature as unattainable ideals to which so called laws of physics are only successive approximations. Now when I say that certain formulations of the laws of atomists of today have a character which is in a certain sense final, this does not fit in with our scheme of successive approximations and it becomes necessary that I offer an explanation. This special character that the atom possesses is the appearance of *whole numbers* [elucidation follows].

We have therefore definite elements in the statements of laws, and there seems to exist a tendency that laws obtain this essential final character when expressed as relations between whole numbers. (Born, p. 2)

Instead of feeling that the old project to substitute the world with concepts was knocked off its platform, the illusion was developed that actually quantum theory secured concepts, it showed us how they arose, even if only in a probabilistic way. It showed the concepts evident in elementary particles were guaranteed by this type of understanding in which we had an insight about the manner of their production. As human beings we had no sense of what happened at that elementary level, what was happening was beyond language, beyond words. But concepts of position, momentum, mass, energy in a probabilistic way through the wave function obtained their character, which then allowed them to be used in all subsequent understandings of how the world worked.

By focusing on measurement, where something appeared, the intervening passage was thereby deftly removed from attention. Bohr used measurement to filter out the very journeying of what happened between the concreteness of observations. The world was necessarily seen through measurement or it did not exist.

Existence wasn't there until you prodded it. So you could hide the nature of existence behind the act of measurement that actually revealed the properties that before had been latent in the experimental setup. His critics said that his explanation was itself concealed in what he was trying to explain.

Bohr would say, 'Well, look at the atom'. The atom was so hidden and concealed, anyone looking at the atom would become confused again. So it took the spotlight away from his explanation. Or if somebody said, 'So what's going on in the atom?' Then Bohr said, 'Well, look at my explanation', which was not really an explanation at all.

Einstein completely rejected what Heisenberg and Bohr were up to. He said:

> The Heisenberg-Bohr tranquilizing philosophy or religion is so delicately contrived that for the time being it provides a gentle pillow for the true believer, from which he cannot be very easily aroused. So let him lie there. (Einstein 1928, in Fine, p. 18)

Einstein intuited that what Bohr was saying was a conjuror's trick that instead of engaging with this upheaval of the fundamentals of science was just a huge hole of concealment. You could say whatever you wanted in that hole of concealment.

Boundary

When a boundary is statically interposed, I say this is my boundary; I believe this; I stand for this; I am of this nation and these are my values; I have decided that before anything else happens. We use a boundary to distance ourselves from the chaotic, from the unknown, the unformed. We say, 'Well everything is safe, I know who I am, and my boundaries are secure; whatever happens out there my position is clear, I know who I am, I know the truth I stand for.' The industrialist and the environmentalist oppose each other with their 'truths'.

It is also the problem that when creating a theory, we do so by defining certain concepts. We make a particular assumption about which boundaries we are going to use, which definitions we are going to approach that reality with. And once we've made those definitions it becomes increasingly hard to see the dynamic that was there before.

What happens in the twentieth century is that the boundary is incorporated as part of the content of what the theory is about.

An example in psychology is the notion of the ego. The ego is declared as part of the psyche. So if you want to know about the psyche, you have to learn about the ego and all the influences, myths, forces, phantoms of the unconscious pulling and pushing the ego. There is a whole story written about the ego. But the definition of 'who am I', 'who is self', is not really part of the content of the psyche. The question of 'I-ness' arises in the setting of the boundary as this dynamically arises in relation to the world. When I relate to the world, then my 'I' becomes clear as who I am through that relation. There is a free relation to the world, in which the 'I' of my boundary is realised.

But in psychology the 'I' of the theory is taken as part of the content of what happens within that boundary of self. We have found a form, an explanation, an order, a description in which somehow we have taken the making of a boundary and made that part of the theory. So it seems like we have understood this form independent of the creative quality of a boundary with the unknown and unordered.

Something quite catastrophic happens because of this. This making of a boundary, this creative quality of a boundary, this primal encountering of the world, between chaos and order, becomes lost. The making of the boundary is incorporated into the theory, into the content.

Measurement in quantum theory is similar. The idea of measurement in the Copenhagen interpretation is put into the theory itself. We hide that vulnerability where order defines itself in relation to disorder. By losing that moment at which the fluidity turns into order, we cannot see any more that there's any indeterminacy within experience. We have taken that point of vulnerability and put it as part of the theory of what we know.

Talking from my own personal experience, there is nothing so debilitating as mistaking our boundary for its content. We cannot live by thinking that our boundary is mixed up with who we are. Because we have this identity that others do not, that is what gives us space to be. That's a completely unviable way to be.

Withdrawal from participation

The detachment of concepts from the process of representation in which they originated, Barfield calls the making of idols. As we have already discussed, this replacement of the world by concepts, led to the substitution of a world of engagement with a pure thought-map of ideas.

> *Idolatry* may be defined as the valuing of images or representations in the wrong way and for the wrong reasons; and an *idol,* as an image so valued. More particularly, idolatry is the effective tendency to abstract the sense-content from the whole representation and seek that for its own sake, transmuting the admired image into a desired object. This tendency seems always to have been latent in original participation. (Barfield, p. 125)

Barfield follows the cultural journey from original participation, withdrawal from participation through to final participation. He connects the process of withdrawal of participation to the experience of Judaism. In Judaism, the very encounter with the synagogue is of a ceremonious tradition that has existed in its own momentum, devoid of any symbols of connection to the natural world. There is concentrated at the centre of this religious and cultural life, a relation to a purely ethereal Divinity, who rules the ethical life in terms of a future to be disclosed in His terms and His terms only. Barfield gives a particularly concise statement of this:

> We cannot resist the conclusion that this detachment from knowledge arose, in the case of the Jews, not so much from any want of any mental alertness as from a positive objection to participation as such. The whole history of the race, from Exodus onwards, is the story of that chronic objection. (Barfield, p. 123)

Striking, as the Jews did, not only at the practice of idolatry, but at the whole religion of the Gentiles, centred round it, their impulse was to destroy, not merely that which participation may become, but participation itself. (Barfield, p. 126)

Maimonides in 1190 Cordoba, where Jews, Christians, and Muslims lived in a transparency of religious ideal, described the exact phenomena in his philosophy of Judaism, where:

... the Divine name ... there is no participation between the Creator and any *thing* else.

All the names of the Creator which are found in books are taken from his works, except one name, the Tetragrammaton [YHWH], which is proper to him, and is therefore called 'the name apart' *(nomen separatum);* because it signifies the substance of the Creator by pure signification, in which there is no participation. His other glorious names do indeed signify by participation, because they are taken from his works. (Maimonides, in Barfield, p. 130)

A withdrawal from participation of the world was set in terms of a Divinity whose authority rested on future spiritual outcome of the role of the Jewish people in moral affairs.

Ultimately these two ideals of science and theology were to clash, in the confusion of the withdrawal from participation, as a secular and religious discovery. Note here the first chapter of Monk's autobiography of Oppenheimer who was to mastermind the effort to create a weapon out of the splitting of the atom.

In *The Life of J. Robert Oppenheimer*, as told by Ray Monk, the personal is the mirror of the scientific.

J. Robert Oppenheimer, his friend Isidor Rabi once remarked, was 'a man who was put together of many bright shining splinters,' who 'never got to be an integrated personality.' What prevented Oppenheimer from being fully integrated, Rabi thought, was his denial of a centrally important part of himself: his Jewishness. As the physicist Felix Bloch once put it, Oppenheimer 'tried to act as if he were not a Jew and succeeded well because he was such a good actor.' And, because he was always acting ('You carried on a charade with him. He lived in a charade,' Rabi once remarked), he lost sight of who he really was. There was a gap and so there was nothing to hold those 'bright shining splinters' together. 'I understood his problem,' Rabi said, and, when asked what that problem was, replied simply: 'Identity.' (Monk, p. 3)

The abstract present is created by sealing the future in a past disguise. That which is possible (the spirit – 'maybe I am special') is attached to an inference from the past (the causal – 'because of my relation to another'). No one can dispute this act of faith that endows phenomena with a secret stamped character putting in inner language a word that takes up all externality; the potential of the future muddles with the causation of the past into a separation of the inner from its manifestation in the outer.

The destiny of the atomic story unfolded through the three key figures: Pauli, Bohr and Heisenberg pictured here beneath. You can see them in their dynamic (notice the pattern of gazes to vacant places) in *Figure 5*.

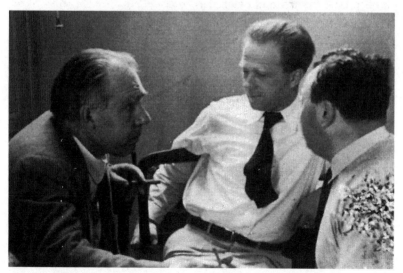

Figure 5. Bohr, Heisenberg and Pauli at the Niels Bohr Institute lunchroom, Copenhagen Conference. June 1936. (Credit: Niels Bohr Archive Copenhagen.)

Pauli was one of the foremost of the quantum theorists, pictured right above. The uneasy solution to the question of what was happening in quantum theory and the riddles that he was grappling with about the nature of reality itself, led to an eruption of existential disturbance. In 1928 a series of traumatic events – the marital betrayal of his father, the suicide of his mother, his own marriage and divorce, triggered a breakdown in his life. He contacted Jung, also in Zurich, where Pauli was a Professor at the University. Under Jung's initial guidance and then friendship, Pauli tuned into the path illuminated in his dreams to a reintegration of the inner with the work he was doing on the outer manifestation of identity in the particle world.

Quantum theory situates the individual particle between the generic potential of the mathematics and the specific actuality of the measurement. Pauli's dreams invert this relationship. The potential is now the living unity, given archetypal form in a specific context of individual event.

The first such dream Pauli had was of world harmony where there was a golden disc with figures of monks going around the disc. There was a great clock in the middle and the monks' job was to keep an eternal time, of eternal balance. And the dream of this world clock, of this harmony existing in the universe, was a first touching of the shore.

Pauli then had a continuing sequence of many hundred dreams in terms of physics, with his dreams being about some beautiful world order.

> Among others, I had the following dream at the time, and it preoccupied me for years. 'A man resembling Einstein is drawing a figure on a board. This was apparently connected with the controversy described [on the foundation of physics] and seemed to contain a sort of response to it from the unconscious. It showed me quantum mechanics and so-called official physics in general as a one-dimensional section of a two-dimensional, more meaningful world, the second dimension of which could be only the unconscious and the archetypes.' (Meier, p .121–22)

There was in quantum theory and in his personal life a dark realm of nothingness, a fitting of properties to meet with expectation, with great power, which one could only approach through light by prodding it, by bringing out the expectations implicit within the system. Pauli's dreams were giving currency to the same conceptual parameters that quantum theory used. But instead of being about nothingness, they were about everything. And you could approach them through darkness, through the unconscious, through accepting that there was this deeper reality, this second, more meaningful dimension to existence.

> What I understand by 'background physics' is the appearance of quantitative terms and concepts from physics in spontaneous fantasies in a qualitative and figurative – i.e. symbolic – sense. I have been familiar with the existence of this phenomenon for about 12 to 13 years from my own personal dreams, which are totally uninfluenced by other people. As examples of the physical terms that can appear as symbols, I should like to list the following without any claim to completeness: wave, electric dipole, thermoelectricity, magnetism, atom, electron shells, atomic nucleus, radioactivity. (Pauli, p. 179)

Pauli saw that if you accepted this revolution in thinking, and followed it to its limit, then you could still talk in the language of quantum physics, you could still talk about dark and light and about this basic interaction with what was there. But instead of it being about the darkness of nothingness, it would become a conversation about the light of everything, the light of the archetype, the light of wholeness.

> Furthermore, my feeling is that *the purely psychological interpretation only apprehends half of the matter. The other half is the revealing of the archetypal basis of the terms actually applied in modern physics.* (Pauli, p. 180 – Pauli's italics)

Physics would come of age when it reflected its findings of the atom back upon the primal need of the age, to guide the inner unity of self-nature into outer expression.

At the time, I was firmly of the opinion that the dreams were a 'misuse' of physical concepts; for that reason I searched frantically for purely psychological interpretations and explanations just so that I could get the physics out of the way; I did my utmost to cling to my conviction that this was all 'just' psychology. But as dreams are compensatory to the conscious attitude, they insisted that the physical terminology should be taken for what it was. This forced me to take it as an essential part of what was being represented. (Pauli, p. 188)

Bohr and Heisenberg

Bohr and Heisenberg, the other pair in the trilogy of discoverers, sticking literally to the theory of an external atom and its implications, travelled on a different path of unclaimed destruction.

Modern science, since Newton, had been working inward through successive skins of understanding, like an onion. It had sought the centre of explanation that would radiate understanding outward to all its layers. Courageously, in the constant dialogue they had between 1924 and 1927, Bohr and Heisenberg reached to grasp the centre in the form of the nucleus of the atom, in what became known as the Copenhagen Interpretation of quantum mechanics. Brilliant terminology such as *complementarity* and *uncertainty* emerged during this time to intellectually encapsulate the onion of science about a purely abstract symbolic centre.

The quantum theory they had developed claimed to have found the ultimate of universal mathematical analysis in the fundamental level of explanation it gave to events at the local level. Science, it was claimed, had transcended the separation between universal law and the local governance of elemental events by a mathematical formulation that accounted for the individual occurrence as a statistical inclusion.

The claim to have found the centre was, in the most terrible way possible, refuted in a visit Heisenberg, the German scientist, made to Bohr, a half-Jewish Dane in occupied Copenhagen in 1941. The world of European culture, their own country's futures and even mankind's survival divided exactly down the line of their quantum theory and the implication that there lay in the division of the nucleus, a weapon capable of destroying any major city and its inhabitants. The nucleus that during their journey of discovery had seemed the safe haven to which they were travelling had cruelly split, leaving the two men as protagonists on either side of the divide.

When Heisenberg and Bohr met in Copenhagen, instead of uniting in this victorious establishing of the atom and the conquest of materialism, they were meeting as enemies, the whole fate of the world hanging between them. They were talking about the splitting of the atom. The potential destruction of humanity hung in the balance of their meeting, on what they'd say to each other. And, yet each of them was an expert at concealment and so in their meeting they couldn't say exactly what they were doing. They couldn't articulate what quantum theory was about. Their whole effort had been in disguising the darkness they had been describing. And so their whole meeting focused more and more and more on this nothingness that was actually at the centre of the atom.

They were both expert at ambiguous talk so that you never quite knew what each of them was saying. Now they were facing the nothingness of non-communication. They couldn't trust each other. They couldn't speak to each other. Instead of having a very ordered world holding of structure, the meeting and the breaking of the atom represented a huge destruction of everything. Everything was bringing the world to a point of destruction, in which there seemed no way to go ahead. They had come into nothingness. They had found the key to unleashing this great energy of nothingness.

In 1956 Heisenberg gave this account of the visit to Copenhagen in 1941:

Under these circumstances we thought a talk with Bohr would be of value. This talk took place on an evening walk in a district near Ny-Carlsberg. Being aware that Bohr was under the surveillance of the German political authorities and that his assertions about me would probably be reported to Germany, I tried to conduct this talk in such a way as to preclude putting my life into immediate danger. This talk probably started with my question as to whether or not it was right for physicists to devote themselves in wartime to the uranium problem – as there was the possibility that progress in this sphere could lead to grave consequences in the technique of the war. Bohr understood the meaning of this question immediately, as I realized from his slightly frightened reaction. He replied as far as I can remember with a counter-question, 'Do you really think that uranium fission could be utilized for the construction of weapons?' I may have replied: 'I know that this is in principle possible, but it would require a terrific technical effort, which, one can only hope, cannot be realized in this war.' Bohr was shocked by my reply, obviously assuming that I had intended to convey to him that Germany had made great progress in the direction of manufacturing atomic weapons. Although I tried subsequently to correct this false impression I probably did not succeed in winning Bohr's complete trust. (Heisenberg 1956)

Bohr's initially unpublished draft reply from his archives:

I listened to this without speaking since a great matter for mankind was at issue in which, despite our personal friendship, we had to be regarded as representatives of two sides engaged in mortal combat. That my silence and gravity, as you write in the letter, could be taken as an expression of shock at your reports that it was possible to make an atomic bomb is a quite peculiar misunderstanding, which must be due to the great tension in your own mind. (Bohr 1956)

The forlorn quest to find intellectually the centre to the onion of explanatory science, is shown in *Figure 6*.

In the meeting of Heisenberg and Bohr, the quest of science turned to this terrible confrontation, in which the destruction of mankind was on the table of discussion. The split between science and the spirit, instead of lying behind in the intellectual roots of scientific inquiry, lay ahead in the spiritual destiny of the meeting between the two men, unable to find trust or understanding.

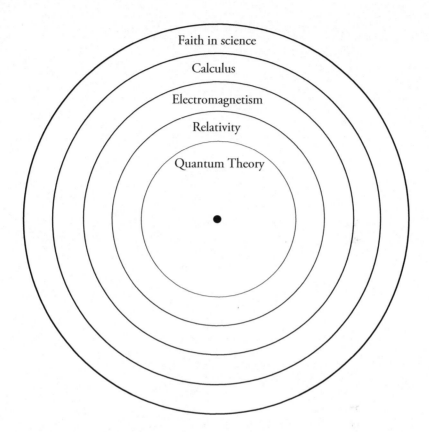

Figure 6. The Onion of Science, working inward, broke in two in the Copenhagen meeting.

The onion split apart into two halves, the journey of science was suddenly going outward to the very edge of destruction from which the only place to safety was the jump to the other half of Creation, as shown in *Figure 7*.

Instead of a concentration, like an onion, where you have layers and layers of science coming to a centre, you had this new challenge to science, which was, how do we understand nothing? How do we understand our relation to wholeness? How do we re-create, like in Pauli's dream of world harmony, a vision of balance and light and creation, out of this point of a complete dead end in nothingness?

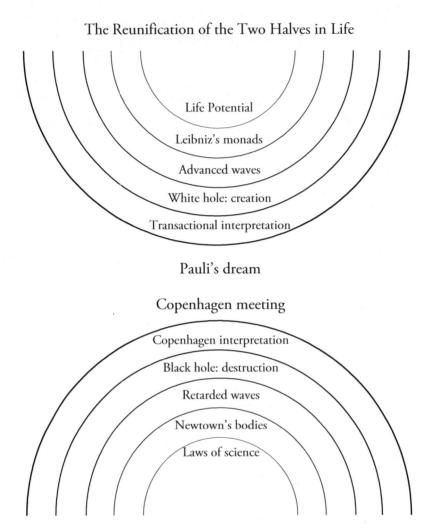

The Reunification of the Two Halves in Life

Life Potential

Leibniz's monads

Advanced waves

White hole: creation

Transactional interpretation

Pauli's dream

Copenhagen meeting

Copenhagen interpretation

Black hole: destruction

Retarded waves

Newtown's bodies

Laws of science

Figure 7. Destruction faces the shore in which the only way forward is the leap to creation.

The only way to go was to somehow jump into this other world of creation. It was to follow Pauli and to re-dream the harmony of the world. Science had reached a complete dead end attempting to describe the world without humanity, without giving expression to anything beyond the material. Science had succeeded in ruling out all human influence at the most elemental of shores where existence comes into being. Somehow one had to get humanity back into the picture.

And many people think today that the split between science and spirit, or the dissociation between man and nature is very trivial. As if all you have to do is explain to people, 'Oh well! You know really the heart is involved, as well as the head.' But this ignores the trough into which science fell and the depth of the breakdown. It fails to see that the act of regeneration isn't going to happen merely by a superficial change in behaviour. We need Pauli's re-dreaming of light and wholeness in order to cross the bridge that this moment brought about, this confluence of people and what they represented at the splitting of the atom.

Henri Bortoft

Bohr's philosophical stance carried the day in early development of the theory. After a while scientists just came to accept the functional efficacy of a theory that time and again showed its outstanding correctness without worrying about the metaphysical implications. David Bohm on the other hand led the attempt after the war to give reality to the realm of potential which he called the implicate order, out of which the explicate order unfolded into manifest existence (Bohm 1980). His assistant Basil Hiley has since done much of the mathematical work to validate this hypothetical idea with a sure mathematical ground (Hiley 2011). But even here the mathematics is hugely complex, difficult to use, and anomalously static.

Henri Bortoft, who also worked with Bohm in Birkbeck, felt from the beginning that the dynamic of the relation of the unnameable whole and the manifest parts had to be brought in as a basic principle of being. Bortoft explored the language we used in describing actual experience. Continual and persistent effort over many years led Bortoft to completely fathom the subtleties of this realm. He also found several related strands in which this exploration had given been given form: hermeneutics, phenomenology, Goethean science.

In the last class he gave in 2012, a few months before his death, Henri told about his early days of working with David Bohm, trying to make sense of the double-slit experiment. As described earlier in this chapter, the photon approaches the two slits as a particle, but produces interference fringes so it must be a wave going through both slits, but what strikes the screen is a particle which should have passed through one slit or another.

Henri Bortoft spent many years trying to give words to this experience, without imposing upon it an understanding from another realm. He said that the breakthrough he had was to understand that for the particle of light there was no separation of slits. The unity was not in the mathematics, but in the very nature of the phenomena. Just as when we describe an experience, the wholeness of the thing we want to express is there at the beginning, but needs the advent of words to give it form. So for the particle, the whole is present to it at the beginning and its particle nature is an expressing of this as a manifest form and movement.

The particle inhabits a dynamic of how the whole moves through into existence of the parts, as an appearing of its quality. One can even, by going on and on with the action of playing the experiment through one's mind, feel the shift between the unity of the whole situation and the actual appearance of the particle as a dynamic that is behind the experimental setup. The theory is an instance whose explanation is not in its causal conceptual parts, as if a mechanism, but in something deeper underlying the phenomena which the experiment illuminates.

Bortoft's journey into the subtleties of this realisation is now published in *Taking Appearance Seriously*. The underlying dynamic of whole/part behind the phenomena is illustrated in each chapter in different instances of language, Goethean science of the plant, philosophy, making meaning, etc. Curiously Bortoft leaves out reference to where the investigation first started in the lab of David Bohm, maybe disappointed with Bohm's reaction when first trying to share these ideas with him.

Conclusion

Bortoft's understanding of wholeness came from the attempt after the war to have a new understanding of what quantum theory was about: to get beyond the point of destruction. It was Pauli's vision of the world clock. Is there, underlying the phenomena of quantum theory, a totally new understanding? Can the mathematics of the atom be one instance of a broader dynamic that has been explored elsewhere in terms of philosophy, language, and meaning? Can the physics allow us to see deeper into a truism about life itself, to discover who we are in a new light? And in order to let wholeness rediscover itself, re-imagine itself, we have to come out from the shadows, wherein Bohr was concealed, to reveal who we are. We have to examine what is wholeness? what is life? all over again.

Bortoft breaks with a science of concealment, that threshold where abstraction encloses existence within a cave of darkness, in which you can hide whatever you want. Bortoft's approach is to discover the process that gets to the light of understanding, the revelation of a dynamic process of whole/part behind nature. He understands wholeness, by arriving at the underlying process, to meet the world in its own terms.

Quantum theory looks at that shore where the world disappears and attention brings structure into being. Bortoft looks at the other shore, seeing wholeness can participate in the journey and show something of essence. So, it is like having two different shores that we are looking at, one this dark shore of fixed existence, and the other, this living understanding of seeing the coming-into-being of something in its essential form. This is what our inquiry is all about, this journey from this incredibly dark and hidden place the world got into, into this very light, and very bright, and very living revelation of our true nature.

Bortoft:

←─────────────────── division/unity ───────────────────→

Chapter 2

Darkness faces light: relativity

Throughout the history of physics, the way the relationship of the individual to the world has been interpreted has determined how mathematics has been turned into the world picture of physics.

The discovery of calculus was accredited to two men simultaneously. Newton (born 1642) in England came to calculus in 1665–67 when living two years in isolation to get away from the Great Plague that was then ravaging England; Leibniz (born 1646) discovered calculus in 1673–75 when also living in isolation, in penury between patronages in Paris.

Newton and Leibniz, with the same mathematics of the calculus came to opposing world pictures through the different bases of relationship of the individual to the world.

Experience teaches me that the world in not black and white, but as Einstein shows, darkness seeds clues for the resolution of light.

Newton

Newton's mathematics was conceived in long periods of isolated study, with an anxiety of exposure and sensitivity to criticism. This meant that he often would not publish his papers for fear of his work being misunderstood. But this changed in publishing the huge two-volume work *Principia Mathematica* in 1686. He then found the panacea to his angst by allowing the reception of this work to mark his inner status. He became Head of the Royal Society, the influential governing body of science in London. This allowed him to push his own work and to aggressively silence critics such as Hooke and also Leibniz.

Edmund Halley's Ode prefacing Newton's great work *Principia Mathematica* begins:

> Lo, for your gaze, the pattern of the skies!
> What balance of the mass, what reckonings
> Divine! Here ponder too the Laws which God,
> Framing the universe, set not aside
> But made the fixed foundations of his work.
> The inmost places of the heavens, now gained,
> Break into view, nor longer hidden is
> The force that turns the furthest orb ...
> Matters that vexed the minds of ancient seers,
> And for our learned doctors often led
> To loud and vain contention, now are seen
> In reason's light, the clouds of ignorance,
> Dispelled at last by science. (Halley, in Newton 1686, p. xlll–xlv)

Here we see the subtle way in which the outside (the Laws of God framing the universe) replaces the inside (the clouds of ignorance). To understand the world on the rational plane, it is necessary to simplify the active play between inside and outside to achieve a mathematical account. In Halley's introduction, this simplification is turned into a philosophical stance. It now seems, reading the introduction, that instead of assuming the outside reality as dominant over the inner of potential for the purpose of this mathematical treatise (a view held by Newton), Halley implies that science has now dispelled the inner, as a final outcome of its calculation.

The replacing of inside with outside reality is thus taken to correspond to an actual discovery of the work rather than a means used to arrive at an understanding.

We see, in the trumpet announcement of Newton's work, the inflated sense of conquest of mystery that was to characterise the next three hundred years of science. Newton went on to be a figure of world renown. *Principia Mathematica* became the defining book of a mathematically rigorous physics that completely changed how science would henceforth be done.

What began as an outer phenomenon to be explored, a causal description for the relation between mass, force and motion, ends up as an internalised reality of how we make use of our cars, plan our flights, transport our goods. The outside becomes inside so completely that we forget that the outside is limited in resources and finite as a receptacle of waste products.

Leibniz

In contrast to this trumpet call to the age of scientific reason, Leibniz's life was completely different. His life exhibits the opposing tendency to substitute inner reality for the outer. He made no attempt to claim outside status. In 1667 he rejected a professorship and chose to work instead for the Baron of Boineburg and to develop his ideas independently. He later worked for three successive Dukes of Hanover, with librarian duties he described as 'onerous, but fairly mundane; general administration, purchase of new books and second-hand libraries, and conventional cataloguing.' (Quoted in Ross.) Thus Leibniz, in contrast to Newton at Cambridge, developed his ideas in the run of everyday life.

Leibniz's understanding of the role that science could play was completely opposed to Newton's. The ambition of his philosophy is set out in *Of Universal Synthesis and Analysis; or, of the Art of Discovery and of Judgement* (c. 1683). In this paper, Leibniz recalls his childhood dissatisfaction with the teaching of the world as deriving from simple concepts. Instead he wanted to realise an internal description of how the world presented itself to experience. This goal of Leibniz was absolutely huge and could be said to be the underlying project of the whole of science that followed on from him. As geometry was able to settle on propositions that elucidated the principles of figures in their relation, so Leibniz was seeking the propositional language that could elucidate a higher justice to existence. In *Of an Organum or Ars Magna of Thinking* (c. 1679), Leibniz identifies the primitive concepts as pure being (God) or nothing. Later he talks about the primitive concepts as the attributes of God.

In other words, Leibniz in complete contrast to Newton, begins by making the inside (God and his attributes) the very basis for the outer exploration!

Leibniz founds his metaphysics on 'the primary free decree of God, namely, always to do that which is most perfect.' Leibniz expressly begins with his own inner experience and then makes this the substitute for everything outside.

Convergence

Calculus gives us a way to understand the motion of a system, according to the rules of interaction between the constituent variables, from one state of the whole ensemble to another. So systems such as weather patterns, structures responsive to outside forces, populations, theoretical equations of physics, all are mapped using the differential equations that are then solved by the method of calculus. So the equations allow one to state how, given an initial set of starting conditions and a context, the system develops out of its own internal propensity to change.

The basic nub of the argument between Newton and Leibniz was the following: Newton still imagined a static universe, bounded by an absolute space and time. So though one understood the equations of change, what they referred to was static things such as billiard balls, or atoms, that were set into motion by external forces but only to be brought back to their natural state of rest.

Leibniz however imagined that the very principle of the universe was one of motion. His very definition of existence, the monad, is paradoxical in this regard. The monad moves to fulfil in its relation to others, the unity of being that it already potentially is at the outset. Thus Leibniz says there is a principle of being between these monads and God, that is motivated to push existence to a unity beyond its current state. Leibniz's aim was to find those propositional laws that allowed existence to develop out of its own tension with itself.

Newton used the method of calculus to gradually remove all partial surrounding influence to arrive at those elemental laws that described motion most primitively. His approximations removed the fluff of possibilities to meet the hard core of material reality, as this obeyed the prompting of forces to instigate motion. He arrived at the most primary definitions of the relation of velocity, mass, force and acceleration, distilled into three simple laws. The laws were by no means intuitively obvious. For instance, a body's natural tendency is to keep to its own speed without the resistance provided by friction. Or that action and reaction are equal and opposite, the principle of a space rocket where the gas thrown out at one end is compensated for by the acceleration of the rocket forward in the opposite direction. So the achievement of Newton was phenomenal.

Leibniz on the other hand wanted to understand through the process of change, the paradoxical foundation that resolved its own dilemma, between part and whole, nothing and God, appetite and realisation, only in the exercise of its being. The underlying dynamic of change was not simply the moving around of bodies as Newton declared, but the moving in the aspiration to wholeness. Change delivered an arrival beyond its own means. Newton was the academician focusing on the laws of physics, Leibniz was more out in the world, curious about the working of the spirit, from the deepest contemplation of experience. Leibniz was described as the last great universal thinker. It is often said that the world was not big enough to house two such geniuses living at the same time.

These two journeys of approximation, one to the solid core of the material essence and the other to the overcoming of a self-contradictory definition of identity in higher unities of meaning, were complementary discoveries from the same leap of mathematical insight.

Newton developed a process of thought distilling the vagaries of circumstance to arrive at the solid centre of material reality from which one could securely proceed in science, one step at a time. Leibniz instead pointed to the living essence of wholeness that could only be reached through experience. Leibniz painted an end goal for an understanding of the world as whole. They thus formed two ends of the whole exercise of science that would follow. Newton's *Principia Mathematica* is the starting point and Leibniz's final propositional calculus is the vast destination.

When taking their two positions to isolated extremes, the mathematics comes to stand for ideal generalities. The caricature of Leibniz was that he was trying to fit the journey of man under God as the end goal of an exploration of mathematical deduction. This was patently absurd! On the other hand Newton inadvertently led science to the impression that conceptual understanding of simple relational systems removed the need for God at all, equally absurd! The only way to go was to believe that the approaches were complementary means of exploring the world. In implementing a technological world on the basis of Newton's mathematics, we would also get to know about the transcendence of wholeness beyond science in Leibniz's vision.

They fit together as the necessary two extremes from which the journey of science is focused. Gradually, in the twentieth century, the new theories of relativity and quantum theory were to loosen 'the Laws which God, Framing the universe, set not aside, but made the fixed foundations of his work'. (Halley's introduction to Newton's *Principia Mathematica*.) Physics moved to a propositional type of inter-dependent logic, more as Leibniz proposed.

The whole of science is visible in the starting and end point of the two approaches.

Divergence

Although Newton wrote up his findings in *On Analysis by Equations with an Infinite Number of Terms* (1669) and *On the Methods of Series and Fluxions* (1671), these were only published much later. The publisher, Collins, knowing privately from Leibniz about his later discovery, tried to avert what he saw as an impending rivalry, by asking Newton to write to Leibniz. In 1676 Newton wrote to Leibniz on his work on fluxions without referring to its content directly. Leibniz wrote back:

> Your letter, dated July 26, contains more, and those more remarkable things in analysis, than many voluminous books published on this subject. Therefore I thank you, Mr. Newton and Mr. Collins, for communicating to me so many curious things.

However it was Leibniz who first published his version of calculus in 1684 in *Acta Eruditorum* (a journal he edited) without mention of Newton.

At this time science and philosophy were in a malleable state, waiting for the big idea that would move society to a more rational base after the turmoil of the Thirty Years War that had ended in 1648. The seventeenth century and start of the eighteenth century were looking for certainty. It was thus the norm for philosophers of the day to present their ideas in the huge canvas of an overarching understanding, without the specialism and reference to other views we have today. One could also remark on the role of the printing press in this dispute, in that arguments were written down as finished versions of ideas that could then compete as to which was the best, rather than being debated orally.

But the future of science was also decided in the competitive spirit in which Newton's and Leibniz's complementary ideas were set at odds by Leibniz's paper. There developed, largely through the meddling of their supporters, a feud as to who was the developer of the calculus. This was resolved rather ominously for the future of scientific integrity by Newton as head of the Royal Society creating his own commission, to settle the manner in his favour. In 1713 the verdict was delivered:

> ... For which reasons we reckon Mr. Newton the first inventor and are of the opinion that Mr. Keill [a supporter of Newton] in asserting the same has been in no way injurious to Mr. Leibniz ...

In 1714, a report was published by the Royal Society in their *Philosophical Transactions* under the heading 'Concerning the dispute between Mr. Leibniz and Dr. Keill, about the Right to the Invention of the Method of Fluxions, by some called the Differential Method.' The report, which has obviously been written by Newton, though anonymously, has 35 pages of description of each step of the dispute around who was originator of the calculus.

The 22nd page of the report, for instance, declares:

> And where it has been represented that the use of the letter o [Newton's terminology] is vulgar, and destroys the advantages of the differential method [referring to Leibniz's terminology used today]; on the contrary the method of fluxions, as used by Mr. Newton, has all the advantages of the differential and some others. (Newton 1714, p. 139)

On the 33rd page, the report makes judgment on behalf of the Royal Society:

> I take the liberty to acquaint him [Leibniz], that by taxing the Royal Society with injustice in giving sentence against him, he has transgressed one of their statutes, which makes it expulsion to defame them. (Newton 1714, p. 150)

There follows then the real reason for Newton's ire which is about the interpretation each is giving to the calculus:

> It must be allowed that these two gentlemen differ very much in philosophy. The one [Newton] proceeds on the evidence arising from experiment and phenomena and stops where evidence is wanting; the other [Leibniz] is taken up with hypotheses and propounds them, not to be examined by experiments, but to be believed without examination ... The one [Newton] teaches that philosophers are to argue from phenomena and experiments to the causes thereof, and thence to causes of those causes, and so on till we come to the first cause; the other [Leibniz] that all actions of the first

cause are miracles, and all these laws impressed on nature by the will of God, are perpetual miracles and occult qualities, and therefore not to be considered in philosophy. (Newton 1714, p. 152–53)

Dimensions

A dimension relates to a freedom that makes equally possible all variations on how it includes and is included by other dimensions. For instance a line is a one-dimensional figure that makes no distinction between the placing of 0-dimensional points along its joining orientation. While the line is itself included freely to make a two-dimensional plane, which itself becomes the free element of a three-dimensional cube. It is then by analogy quite possible to continue this process of embedding say a three-dimensional space in a fourth dimension. The three-dimensional figure, the cube, is included freely as the element of a fourth dimensional orientation. The three dimensions in which we operate in everyday life, are drawn from a higher dimensional reality.

Given we are free human beings, there are different ways we can put together these dimensions of reality. Either we can start with what we know, on which to found everything we experience, (as Newton); or we can start off with our relationship to the unknown, distilling to an essence of everything we can become (as Leibniz). Neither is a better way of seeing the world, just a reinterpretation of the dimensional foundation of freedom in which structure rests. What is more, the same mathematics relating the dimensions of space and time to existence can apply to both approaches, as calculus was the basis of both persons' physics.

Newton's physics was from the perspective of three-dimensional bodies in motion, Leibniz on the relation of our three-dimensional world to the unities of the infinite. There is a necessary ambiguity in how we participate in the world, either as a player in the dimensioned world of space and time, or as the fulcrum about which exploratory journeys come together in the resolution of meaning.

There are two different questions to be considered here. One is the nature of the three dimensional reality and its functional characteristic. Newton discovered that the basic reference of three-dimensional space is the constancy of the motions within it. The other is the relation of change to dimensions imagined through paradox. Change orders itself in dimensions of the finite in paradoxical juxtaposition with the infinite. The paradox of experience is how the totality of possibility of our relation to the world finds finite form in which to be explored. For Leibniz the dimensions of experience derive from the ability to live the whole through partial event.

In the dimensions of paradox, considered by phenomenological philosophers such as Kierkegaard and Bortoft, experience wholly mediates the two sides of the finite and the infinite. A power in Kierkegaard's language (the infinite unity) transparently establishes the ground of separation (the finite world of things). Bortoft called this going 'upstream', giving ground to the world of dichotomy in discovering the active power that establishes the relation, as drawn in *Figure 8*.

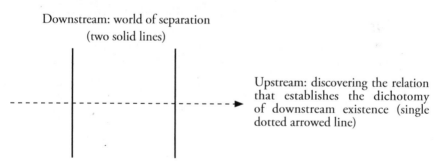

Downstream: world of separation
(two solid lines)

Upstream: discovering the relation that establishes the dichotomy of downstream existence (single dotted arrowed line)

Figure 8. Ground of separation in the line of relation.

In experience alone, can we go beyond the world of separation. As long as we stay in the world of analysis, there always exists a dichotomy raising its head somewhere, as in the different interpretations of calculus by Newton and Leibniz. What phenomenology does is discover the creative formative relation out of which the dichotomy of interpretations originates. This formative relation is not another part of the system of separation. The ground it provides is a resolution coming from outside the limited finite system itself. It has to be encountered.

The world happens between the matter of Newton and the meaning of Leibniz. When we see the world through the lens of separation then these two viewpoints inevitably present their polar opposition. But in its very happening, the world establishes ground between matter and meaning for some higher unity to present itself. By going into the happening, which is after all what physics wants to describe, we arrive at Einstein's theory of light as the ground between Newton and Leibniz.

Existences and completions

The complementary ways of seeing the world that developed from the mathematics of calculus show two perspectives.

For Newton, calculus applies to inert matter and describes how forces interact with bodies essentially resistant to change. The world would like to be at rest and has to be brought out of that state by forces applying change.

For Leibniz, calculus applies to the innate capacity for change. It describes how interactions between potentials bring about the archetypal unity of experience, transcendent over change. The world is in change, seeking completion and coming to rest in the monad.

The basic principle of how change leads to stability is interpreted in two ways. In interpreting the world through the elemental notions of secure fundamentals such as atoms and genes, one will see the higher level structure of cosmos and organisms defined in relation to their lower parts (Newton). However, in seeing the world as an innate potential to realise the higher, then the dynamic of participation will take one up into the monads, defined in self-relation as intrinsic wholes (Leibniz).

The potential for individual relation to the world recognises both: the thing in part relation to other things and a soul in whole relation to itself, through the same mathematics. There is simply the twist that the *definition of things* is through a relation to their component parts and the *wholeness of soul* is a self-relation. A thing is itself locally through the parts that define it. The soul is in journey to establish itself in self-relation as whole.

We find the choice of interpretation to be present at every stage of life. And that we shall go forward by recognising and relating these two ways of seeing at every stage of physics, right up to the definition of the atom. We see the world and either turn the key of the thing in terms of its parts or we turn the key the other way and we see the world as relating through its wholeness

And, what changes the way we turn the key, is our commitment to how we see the world. The question we ask the world is going to decide if we see things in relation to their parts, or the soul-essence of potentials relating to themselves.

Freeing the potential

At Oxford, during my mathematics degree, looking to the career offers, sometimes a totally different potential seemed to call to me. This potential could be aroused by the aesthetic of someone playing a guitar and seemed to call to me to take a completely different path.

In the act of putting the key through the locked desk, as described in the introduction, I had sought to establish existence along the lines of a certainty. This feeling of potential, on the contrary, yearned to explore everything that concepts did not seal into fixity. I wanted to understand change, not as it was captured in finite description of itself, but as it led to a unity beyond itself. At the end of passing my finals I bought a travelling bicycle and set off towards Africa. The journey, from all the different standpoints I was surrounded by, sought to find unity.

The journey is described to communicate the subtle thing that happens with experience, when one pays attention to it in its happening.

Travelling gives illustration of the character of experience. When travelling, the journey at the start is driven out of the tenacity of will seeking expression. But gradually a form to the journey begins to emerge. Various circumstances come together in giving the voyage a character. To some extent this character is determined by the lands travelled through and the people encountered. But in other ways the character is unique to itself. Gradually over time, the journey itself begins to assert its own choices as to how its whole potential needs to be served. I become selfless, simply responding to this greater endeavour with all its risk and uncertainty. Then at a given culmination of adventure, the voyage flowers into the realisation of its whole nature. Something happens. The travelling becomes an achieving that names itself through the adventures along the way.

My trip began, at age 21, as I finished my finals at Oxford with no fanfare or expectation whatsoever. It began in a rainy Rotterdam one July, with the person whose idea it had been in the first place to bicycle and then hitch with me to Africa giving me a send-off, as he had no money to accompany me. I had very little experience of cycling anywhere and to spend the day on the bike seemed a challenge enough. So I set off on this trip, with six months ahead of me before starting a computer programming job, in an almost laughably mundane way.

At first it was raining and my efforts were judged by passers-by, especially in France, from a mechanistic perspective. Why was this daft Englishman expending so much effort to carry himself, tent and belongings through the rain?

Subtle things happen with experience. Gradually, as one kilometre followed another, as one hill was ridden up after another, as one day was navigated and then another, a pattern of experience laid itself down that began to identify itself as its own story. Even though the effort of riding the bike seemed at times quite a pointless activity, the repeat of the activity gradually distinguished to me and the world that something was going on here. Time was not simply taking each day as some new beginning of labour as one might feel sometimes in an office, but was trying to make sense of what I was doing as an activity in the world.

Even though my effort was very much rooted in the application of force to make the bike and belongings move as Newton had understood the world, a quality of wholeness, the monad of Leibniz wrote itself down through these moments.

Especially when reaching Spain with the sun coming out, there was a greater understanding of the more subtle whole qualities of life. I told my stories in exchange for milk from the cow and a place to pitch my tent on the farm.

At Algeciras on the tip of Spain, I sold the bike and continued by ferry to North Africa.

The experience of crossing the Sahara was a rite of passage, in all its beauty. The hard work of keeping the Mercedes car and van free from the desert sand further differentiated my being in the very enterprise I was engaged with. The greater the severity of the test, the more the experience wove the uniqueness of its own special type of resolution. We arrived at Arlit and Agadez, on the edge of the desert, full of native African smiles and charisma.

Where does humanity come from, suddenly, surprisingly, out of the desert, singing back his song to the silence? In contrasting the way of song of simple African villages with the technological frenzy I had left behind, I wondered in my diary:

> Western man is so set on his blind course that carries all along regardless of direction or purpose, could it be that he has passed by, in his rush, some nobler life of altogether greater value and purpose? (Franses 1980)

There exists a dimension of existence that balances its potential in whole opposition to everything we know by division. If I try to follow this dimension to the bedrock of understanding, there is nothing; however if I allow the question its freedom, then some unity settles into the midst of my experiencing. Another such question hints that the universe itself might possess the same quality of arising out of the paradox with its own finite material statement.

> If evolution was such that the senses, in particular sight, had never been developed in life, then, in what sense would the universe, all infinity of it, exist? (Franses 1980)

Travelling whole

The questions about the nature of existence seemed to have all been rendered academic when I reached Bangui, the capital of Central African Republic. The guidebook I was shown announced that the road to Sudan I was hoping to take was almost unfrequented by vehicles (less than one car a month was their estimate of traffic frequency along the later stages!).

A teacher, Mark, heading from Nigeria to the UK, seemed to provide a providential intervention.

> 'We've been to see the Encounter Overland group and they've just received a telex from London advising them to cross Central Africa to Sudan and then to go down to Kenya to avoid the trouble in Uganda. We're hoping to tag along with them.'

My suggestion that I travelled with them was at first entertained. However this dream of outward safety dissolved in the fears about the state of the route.

> 'Look,' said Mark, 'I'm not sure we can actually take you. My wife's very worried about the children and then there's the inconvenience.'

I realised at last that, whatever the problems I might otherwise face getting to Sudan, a lift with Mark was not the answer. The only way I could count on a lift was to get most of the way on my own. It was obvious what I had to do – get to the road and start hitching.

After a lift to Bambari, I was passed by the Encounter Overland truck and Mark the teacher, heading to Sudan. But another encounter brought me into contact with a UNICEF inspector delivering supplies to schools, and though at first appearing full, I eventually bargained a place on the back with some other Africans heading home.

With the dense forest/jungle all around and seemingly nowhere near any human habitation, it was a surprise when we saw a truck stopped up ahead. As we approached I suddenly recognised the unmistakable blue and orange of Encounter Overland and they told me how they had driven into a tree that had fallen across the road and wrecked the front axle.

The next day we came across Mark and his family also stopped on the road with technical problems to do with the trailer.

There are moments when one follows one's heart that it seems one triumphs over the whole world, just by the exuberance of one's effort, as if the world is taken from its static tracks and lifted up into the air. In testing my being against the world, it was the chance of encounter that temporarily triumphed over the secure way of mechanism.

It was in this mood we arrived in the village of Zemio. We were held up partly by there being no petrol to take us further.

In Zemio we received a tremendous welcome, the whole village seeming to know someone in one of the two cars. It was a bit of an anti-climax though to hear we would have to wait 'two or three days' before moving on.

From my diary:

It is a Monday morning and all over the world there are those waking up and cursing the start of another working week, eating a silent breakfast cooked by the bleary-eyed wife, taking their general annoyance out on the children for spilling the sugar.

In Zemio at 5.30 am, the final respects were being paid to a young woman who had died tragically three days ago when her thatched hut had been set on fire by lightning and she had been burned alive. By 6.00 am the gathering breaks up and the sad faces begin to lighten. A family sits around a fire laughing at a young kid of no more than three years old, dancing in time to the music of a radio. Elsewhere other groups sit, chatting and laughing, around the fires, over which breakfast is cooking. No one is worrying about what's going to happen that day or that week or any future time; what's going to happen is life – work to do, friends to talk and laugh with, kids to raise, love and pain.

Blue sky, mud brick thatched huts, pots over the stick fires, chairs to sit around. No toaster, no dishwasher, no oven, no kettle. No powdered orange juice, boxed corn flakes, canned beans, packaged sausages. No clean white shirt, pressed suit, matching tie. No car to drive to the station, no train to squeeze into on the way to town, no office to go to work in. No mahogany banisters to be polished, no silver tea-set to be shined, no decorative plants to be watered. All the complicated masonry of modern, high-pressured skyscraper existence is missing and what is left – peace, contentedness and happiness which the masonry had only served to bury. (Franses 1980)

The paradoxical juxtaposition of death and life, anguish and joy, emptiness and event is the living basis on which the dignity of the village builds. The struggle in my own soul relaxed into this simultaneous allowance of opposites as a basis for life. On the one hand I was absurdly stuck without transport in the middle of nowhere and would sink into periods of despair at my situation. On the other hand, there would be moments of pure jubilation at the freeing of my spirit to exist in its own light of potential.

From the perspective of paradox, one of the most noticeable things is the different relation to light. When one knows despair, darkness and hopelessness, then light, when it arises, is something that appears playfully in its own quality, dwelling in its own order, asking to be extended further. Light unifies and connects different possibilities together, is transparent to different approaches, is able to unify existence separated over space and time.

Between the times of the day set aside for worrying about the logistical situation, my being would know light as a transparent quality resolving all the paradoxical qualities of my situation. At these moments, the separation of space and time would dissolve into a unity of experience that joined together many disparate pieces of knowledge into a whole sense.

On the sixth day in Zemio, still waiting for the petrol to arrive, I wondered whether it would be possible to translate the paradoxical approach to life, and in particular the experience of darkness and light, to Europe, and the technological world to which I was returning. Although I did not know it then, I was intuiting the ground Einstein had created in basing physics on the paradoxical place given to experience by light.

Einstein's light

As Bohr struggled with the darkness of the atom, figuratively and metaphorically, Einstein floated over a landscape illuminated by light.

Einstein was the traveller who wanted to understand the nature of journey that allowed one to appreciate the unique quality of the world one was moving through. Given that one has a moving frame of reference, one can assume nothing about the identity of the world. How does it happen that one encounters in this motion, a quality of existence that communicates a whole integrity? Einstein's first insight was that the traveller on a light beam would know no distance or passage of time. His journey would be inseparable from its starting point to its end. From this insight, he explored the hypothesis that not only did light travel in space and time but that space and time were dynamically determined by the observer's relation to light!

Einstein's assumption was that the speed of light was invariant to any relative speed of observation. Light transcended partiality as its speed in vacuum was constant everywhere.

Einstein then asked how do space and time, as arenas of possibility, lend their medium of happening for light to manifest? What is the necessary dance that maintains the integrity of a spatial temporal world to answer all paths of inquiry about the universe's nature?

With one clue, the fact that light is an independent instrument of orientation, an equation of change between mass, space and time is derived, for which the universe is a unique and necessary solution.

When we let go of all fixed assumptions about the construction of the universe, the dynamical interplay of potentials still deliver their own order. Though everything is up in the air, structure finds a single way to offer a unique course on which to deliver existence.

As Reichenbach says:

> The system of causal ordering relations, independent of any metric, presents therefore the most general type of physical geometry. If rigidity and uniformity were to disappear, the causal chain would still remain as a type of order. Although everything is in continuous flux, there is a structure discernible in this flux. It is striated and can be resolved into chains that define a strict topological order.

Einstein was above all adept at putting concepts to use, not as literal translations about the world, but in the way a painter might use his palette to further illuminate a subject. His work was totally revolutionary, for in giving up the literal nature of concepts as descriptions, the dynamic interplay of terms gave the order of temporal relatedness we recognise as world. Physics underwent a sea change, from an understanding of Newton that the natural state of the world was the maintaining of an established order, to the more Leibniz-like understanding that what was natural in the universe was a participation in a dynamic.

> While we thus see in the causal theory of space and time the philosophical result of the theory of relativity, we wish to point out that this idea of a causal space-time order was conceived long before the advent of the theory of relativity. It was none other than Leibniz who developed in his *Initia rerum mathematicorum metaphysica* the basic ideas of this conception. It is the more remarkable that Leibniz, this genuine philosopher, was able to understand the nature of scientific knowledge to such an extent that, two hundred years later, a new development of physics and an analysis of its philosophic foundations confirmed his views. (Reichenbach, p. 268–69)

The relation to light becomes, with Einstein, the propositional foundation by which the dynamic of the world realises its substance. Einstein made the crucial step that mediated the contemplation of God with the science arguing the world from nothing. Einstein found the middle C that allowed a scale to play between the two extremes. (c in physics is the symbol for the constant speed of light.)

The closeness of Einstein's vision to the spiritual vessel of Leibniz can be found in the relation of Einstein's physics to the philosophy of Martin Buber. At the same time as Einstein was relating through a map of symmetry the qualities of the natural world that formed through their interactions, Martin Buber was doing something similar for the spirit in his book *I and Thou*.

Religion had fallen with science in similarly mistaking law for definitions: soul, spirit, scripture had become mere labels taking mankind from the freedom of their ethical responsibility to a mechanical obedience to tradition. Martin Buber showed that the choice of the word-pair *I-It* or *I-You* actually takes place *before* one constructs either a thing or a soul-world. Every labelling of the world, turning undifferentiated nature into a thing of objectivity, is, by its very definition, making a lifeless map. This map however excludes the symmetry of the living encounter, in which every being of the world finds its own place.

> The world is twofold for man in accordance with his twofold attitude.
> The attitude of man is twofold in accordance with the two basic words he can speak.
> The basic words are not single words but word pairs.
> One basic word is the word pair I-You.
> The other basic word is the word-pair I-It; but the basic word is not changed when He or She takes the place of it.
> Thus the I of man is twofold.
> For the I of the basic word I-You is different from that of the basic word I-It. (Buber 1970, p. 53)

And yet paradoxically there is a fundamental difference between the two men, in which the potential unity of their positions is hidden.

In the hands of Einstein, the physics of the word-pair *I-It* produces its own stunning mathematical beauty, elucidating a single relation that stands for everything that occurs in the phenomenal universe of spatial and temporal occurrence. When Einstein takes into account the act of observation turning the wholeness of the world into a space-time coordination, the result is a brilliant insight into the completely wondrous nature of the physical world. For Einstein this sensational arrival is enough.

For Buber however it is the word-pair *I-You* alone, the insistence on meeting existence in its own spontaneous appearing, holding to the one-off uniqueness of every being, which leads into the address of the Divine. Buber sees the world of *I-It* as a shadow continually following behind the world of immediate encounter of *I-You*, to conceptualise experience out of an analytical folly. *I-It* is an out of control abstraction upon abstraction.

Where Einstein oversees in one glance the symmetry of the many, Buber pierces through with the unique and unrepeatable encounter with the essence of the One. Light in Einstein's theory plays a role that absolutely transcends the reduction of experience into a simple aggregation of things. The constant speed of light, regardless of the speed of the observer, redeems the unity that Buber sees lost in the choice of *I-It*.

Light's unity

The finite speed of light determines that any physical theory is going to be an interpretation of a relation, individual to world. The finite perspectives of space and time achieve their significance in relation to the unity of light. One first needs light to then give significance to the finite world one is investigating.

We shouldn't try to put light into an existing structure, but see the world as it comes into existence through light. Instead of trying to fit light into existing physics and the description of the world, Einstein argued, we should get physics to look at the world as it comes into existence through light.

In *The Speed of Light,* David Grandy writes:

This then is the explanation for light speed constancy: too close to home or informative of our nature to be objectified, light is not distinct from our experience of light. When we therefore measure the speed of light, we are measuring something that remains in play, something that keeps moving with us on our experiential way. Put differently, to measure the speed of light is to measure something about the way we ourselves are measured or blended into the cosmos, and that universal blending pre-decides our measurement of light's speed in favour of a universal or constant value. This is a constant of nature inclusive of our nature. (Grandy, p. 8–9)

Light is something that gives the world character, even though we cannot say what light is. The world isn't an absolute world that's being seen, but a world given by light. The thing that triggered Einstein into his theory was to imagine someone travelling on a beam of light.

Bernard Haisch remarks:

If it is the underlying realm of light that is the fundamental reality propping up our physical universe, let us ask how the universe of space and time would appear from the perspective of a beam of light. The laws of relativity are clear on this point. If you could ride a beam of light as an observer, all of space would shrink to a point, and all of time would collapse to an instant. In the reference frame of light there is no space and time. If we look up at the Andromeda Galaxy in the night sky, we see light that from our point of view took 2 million years to traverse that vast distance of space. But to a beam of light radiating from some star in the Andromeda galaxy, the transmission from its point of origin to our eye was instantaneous. (Haisch, p. 30–31)

This led Einstein to his theory of relativity. He imagined being on a beam of light and the realisation that you'd experience no distance, no separation, propelled him to see that all finite perspective of space and time was necessarily related to this absolute nature of light, as non-separate.

Because light creates the world of separation, when you ride at the speed of light, there is no separation, no space, no time. Light is beyond separation and creates the world of separation. The world of separation is brought into being by non-separable light. Light establishes the ground in which separation is revealed. The non-separable brings into relation the world of separation.

So, light has this quality of being both part of the absolute and indivisible and yet, at the same time, part of the structure of separation. It is paradoxical that you can't separate light, you can't divide it, you can't even see light, yet it is what presents to us a world of separation, of things.

Light illuminates the world of separation. Light makes manifest what is beyond separation. Light makes the separation appear as meaning.

> By means of light-mediated images we are able to have visual experience of things not materially present. As a presenting medium, however, light does not yield to such experience. To bring off vision of other things, light must be 'the 'letting-appear' that does not itself appear.' It must also be the 'letting appear' that gives rise to visual experiences not spatially and temporally coincidental with the things experienced. In letting all this appear or happen, light is not party to any of it. Its lack of appearance, its clearness or invisibility, keeps it fully present rather than imaginistically divided off into places we call 'there' and 'then.' (Grandy, p. 56–57)

If we look out at a tree, then the tree is a long way away, but light brings us the tree. The idea of space, separation, of us being here and the tree being there, is transmitted by the fact that the light doesn't distinguish here and there. Light creates for us an understanding of separation between us and the tree. Light is not playing that game of separation. Light is simply joining the image of the tree into our eye and revealing the ground of separation, as the distance between us and the tree.

Light mediates between the worlds of wholeness and the world of separation. As Bortoft puts it:

When it comes to science I think paradox is to be expected. Think of light in the special theory of relativity. It is a consequence of the universal constancy of the measured speed of light that light itself is not subject to the space-time separation which is characteristic of material bodies. Light itself is before separation, and it is a consequence of the null-interval that the universe for light is an intensive point including all within itself. To the logic of solid bodies, for which separability is a defining characteristic, such non-separability is highly paradoxical to put it mildly. But now imagine a being of light. For such a light-being the world of bodies would be impossible to imagine, and the idea of separability would be highly paradoxical. So if we say that the behaviour of light is paradoxical, we should not imagine that this paradoxicality is somehow intrinsic to light itself. Non-separability, in whatever form it takes, will always seem paradoxical to us in the world of bodies where separability is the major characteristic. (Bortoft 2010, p. 34)

When we try to divide the world into Newton's separation and Leibniz's non-separation, we find we get into this paradoxical situation where both have to be true, the separation of Newton and the wholeness of Leibniz. And we don't arrive at this paradox by abandoning logic. It's by going through logic, by going through Einstein and his understanding about how the world must be constituted, that we find this counterpoint between wholeness and structure.

Wholeness and structure

The nature of a journey is to leave open the paradoxical relation for the events themselves to evidence a meaning that is born through them. The paradoxical relation of wholeness to parts is left open by experience to allow the meaning that is discovered at the end to enter in through the journey, as a sense given to the individual occurrences upon the way. To really know the wonder of light, we also have to experience the eclipse of darkness.

The book *The Ritual Process, Structure and Anti-structure* by anthropologist Victor Turner reflects on the process to reacquaint the individuals with the ancient holding unity that is reborn through the confrontation with darkness. An example is children in passing into adolescence from childhood. The young villagers are separated from all reference points of the old cultural norm in order to be given their new role in the mythological story as adults. But also transitions into marriage, parenthood and elder of the village are discussed.

Turner writes:

> During the intervening 'liminal' period, the characteristics of the ritual subject (the 'passenger') are ambiguous; he passes through a cultural realm that has few or none of the attributes of the past or coming state.

> It is as though they [the liminal entities] are being reduced or ground down to a uniform condition to be fashioned anew and endowed with additional powers to enable them to cope with their new station in life. (Turner, p. 94–95)

The signs I met along the way of my setting out from Bangui, the overtaking of the two foreign vehicles, the dark-light nature of my stay in Zemio, are the fluid putty of events that form into the whole unity of journey. These signs are not fixed as elements of a theory that could be repeated; they offer themselves dynamically to establish unity from randomness. In setting out with determination into the unknown, the signs orient the transformation of experience into a new stage of life.

Paradoxically, in setting out without order, a structure reveals itself as necessary to deliver a dynamically realised meaning.

The dark realm is home to potentials of the spirit, which have, innate in their being, a quality of unity through experience. Although paradoxically such potentials embark on a wild diversity of disorganised adventure, these trials lead to a quality that knows itself in reference to a unity. The potential is signified in the context of the world through which meaning is made.

'Participator' is the incontrovertible new concept given by quantum mechanics; it strikes down the term 'observer' of classical theory, the man who stands safely behind the thick glass wall and watches what goes on without taking part. It can't be done, quantum mechanics says. Even with the lowly electron one must participate before one can give any meaning whatsoever to its position or its momentum. Is this firmly established result the tiny tip of a giant iceberg? Does the universe derive its meaning from 'participation'? Are we destined to return to the great concept of Leibniz, of 'preestablished harmony' ('Leibniz logic loop'), before we can make the next great advance? (Misner *et al.*, p. 1217)

Walk

The notion of order as a universal certainty I could count on was fundamentally eroded in the wait for transport in the village of Zemio.

The Encounter Overland and the teacher passed through the village, their technological problems fixed. The petrol arrived and we travelled to Obo, the car's destination. It was clear I would have to walk 110 kilometres in the next three days through the lion-filled jungle, not knowing if I would be eaten or would walk through safely.

On the way to Obo my fellow passengers mocked my overt nervousness. Pointing out the graves with a stone hand on top by the side of the road, they explained to my tourist innocence:

'See, Philip, this is the hand of a tourist who was eaten by a lion!'

Arriving in Obo, going out in search of palm wine brought me in contact with Joseph who was walking to the border the next day.

'Are there problems on the route with animals?' I asked.

'Not often!'

In everything that had brought me to this walk, my childhood, Judaism, suburban surrounds alienating me from nature, concrete encroachment of civilisation, schooling, family entanglements, university, I had sought a comforting certainty to meet with outside expectation. The walk took me beyond the edge of all the implications of these separate issues. It took me from an arrangement of concepts in which I was neutrally implicated, into a pattern in the world in which I was actively called. In surrendering into this crucible of possibility, back to my own origins, light showed itself as a promise, holding the order of what could be realised through the challenge of event. Light took the fragments of experience, taken out of their old matrix of safety and imagined for them a different resolution, born out of the nature of worldly challenge.

The paradox of darkness and light that in Zemio I could still ponder on from a distance now engulfed me directly. The dimensions of space and time were no longer guaranteed to stretch the 110 kilometres and three days to my destination.

The space-time landscape

The intimate relation of space and time with matter was explored by Einstein in his general theory of relativity. This considers how space and time are bent to incorporate the effects of matter, in such a way that bodies in the space-time landscape appear to be drawn by a gravitational pull. Einstein thus fully explicates the balance between space and time on the one hand and matter on the other.

Einstein shows how space and time respond to matter (and matter to space and time). His differential equation (relying on calculus) shows that gravity, the inclination of space and time towards the centre of a large body such as the earth, is exactly accounted for by the inner distortion to the space time experience of oneself as a falling body, in relation to the space time perspective of the earth.

However in 1912 Einstein couldn't even believe his own theory. Space and time now took on their unique relative aspect according to the mass in their vicinity. Where there was *no mass* present, the equation was unable to predict any exact solution. This was known as 'the Hole Argument'. The light that gave order to the universe could only make a local interpretation of the meaning of space, time and matter. Either space-time-matter all arose together in a neat definition of their collective property of existence, or the equations showed nothing at all.

Einstein struggled with this the rest of his life. Nearing his death in 1952, he wrote:

> On the basis of the general theory of relativity, space as opposed to 'what fills space' has no separate existence. There is no such thing as an empty space, i.e., a space without [a gravitational] field. Space-time does not claim existence on its own, but only as a structural quality of the field. (Einstein 1952, p. 155)

This contradicted how Newton defined space and time *(Definition 8 of the Scholium to Principia Mathematica)*:

> Absolute, true and mathematical time, of itself and from its own nature, flows equably without relation to anything external, and by another name is called duration.

> Absolute space, in its own nature without relation to anything external, remains always similar and immovable. (Newton 1686)

It however agreed with Leibniz who wrote:

> As for my own opinion, I have said more than once, that I hold *space* to be something *merely relative*, as *time* is; that I hold it to be an order *of coexistences*, as time is an order of successions. For space denotes, in terms of possibility, *an order* of things which exist at the same time, considered as existing *together*, without enquiring into their manner of existing. And when many things are seen together, one perceives *that order of things among themselves*. (Leibniz & Clarke, 1715)

In other words space and time have no intrinsic meaning as being reserved for events that happen materially in an objective world. The understanding of relativity applies also to the quality of experience in how time develops the meaning of its content. An inner journey develops its story in a rhythm of unfolding that discovers coherence as it develops. Space and time happen in developing our own stories, however we do that. The stories make themselves embodied actualities in their own temporal order with their own energetic resolution.

Arrival

I arrived at the border post into Saci Yebu where the border guards busy playing dominoes brought me continual glasses of water on the news that I had walked from Sudan. The feeling which greeted my sitting on the chair sipping water at Saci-Yebu, was a presence in something more weighty than just the finishing of the walk.

There is a space of possibility that we enter and that space of possibility is the moving thing that delivers a resolution. That resolution has an illumination to it. When we realise the illumination, time comes into being.

It is not that space and time are there at the beginning. There is in the beginning a space of possibility and that is what coheres through an illumination. Then we can say there is time and an account may be told of what happened.

> I hope it is also possible to appreciate the great satisfaction I feel at the end of the trip. I *know* these six months have in no way been wasted – there's been so much that before was outside the sphere of my thought, that without the pressures of modern living and with the stimuli of travelling, I have been able to perceive and think about at length. Ideas from nowhere, which in various high moments I was able to comprehend in great clarity and that are now written down in black and white. I feel very much 'richer' for them and that I have made some sort of step forward on life's path.

> In the same way I look at my new gained knowledge about the world that has now become that much less of an unimaginably vast and largely unknown planet; just looking at Africa on a map of the

world, it is no longer the geography subject that it once was, but it is people, places and memories...

Travelling has been like a teacher that has made me aware and wise about so much that modern living rather leaves hidden in a corner. (Franses 1980)

We began the chapter with two interpretations of the relation of individual to world, Newton and Leibniz, in dispute with each other over their philosophical stances. In entering the quality of experience, I learn however to hold paradox. Instead of starting with dimensions and trying to fill them with actual defined objects, paradox allows us to journey in non-definition until the resolution gives to the paradoxical dimensions the roundedness of completion. Einstein uses the paradox of non-separable quality determining the world of separation, to hold the question of the nature of the universe in the paradox of darkness and light.

Einstein:

$$\longleftarrow \text{_____ darkness/light _____} \longrightarrow$$

Chapter 3

Past faces future: electromagnetism

In the last chapter we suggested that Newton and Leibniz marked two ends of a journey. There was the belief from both men that they had understood the direction which the new exact mathematical science would take, and yet their intuitions took them in opposition directions.

The world, to be understandable to itself, has to possess an ambiguity, in which each event contains a reflection of itself, allowing its meaning to be grasped in the continuity of all actions.

An electromagnetic wave is always generating its own order out of itself and so is a prime candidate for mathematical description. Maxwell's equations of electromagnetism that united electricity and magnetism in 1861 had two possible solutions.

In one solution, the electromagnetic wave (such as light) travelled with a huge but finite speed forward in time. This solution we could call *Light in Time*.

The wave can be stated mathematically because its future is totally identified from its past. There is continually a reflection of what the world is, through what it has been. But this self-reliance on its own generative essence can be turned around the other way. Imagine now a wave where the past understands itself by the freedom left open for the future to provide a defining meaning.

This perspective gave another hidden solution to Maxwell's equations of a wave moving ahead of time, advanced from time. One could call this solution *Time in Light*.

This solution by definition falls outside the scope of understanding through physics. Physics is the study of happenings in time. But in this reverse solution, time is embedded in something beyond itself. Time develops events that may announce their coherence in light. The dependence of events on their self-generating order leads to a resolution, whole unto itself, the monad of Leibniz.

Ambiguity

A string theorist will say there are eleven dimensions to the world and they will have an argument with another string theorist saying there are ten dimensions and they will have an argument with somebody who believes in relativity who says there are four dimensions and they will have an argument with a quantum theorist who says the world is not a continuity, it happens in discontinuous leaps. And all these arguments take place within the scientific community. But imagine an artist, who will say, 'I don't care for science, science is completely irrelevant, what you have to do is feel the world and be creative.' Or a theologian will say, 'You have to follow the Word that has been given'. All these different perspectives in the world are allowed to coexist, each in their own right and their own definition, but each to some extent contradicting the other. There is a primary ambiguity in the world that allows the world to be interpreted in different ways according to how you come to it. The world is, in this sense, seen more as a question that according to what you ask can give different answers.

What is this thing called ambiguity? Some definitions of what people understand by ambiguity: approximation, things that are not clear, unknown, paradoxical, contradictory, confusing, having more than one meaning or interpretation that only context can decide.

Originally science had been called Natural Philosophy and the people who did science were these quirky people who either had an enormous amount of money or a rich patron and could spend their time devoted to looking at the heavens through a telescope or looking at electricity or, like Newton, dabbling in alchemy and working at science according to his own plan. And when it was this quirky thing called Natural Philosophy, discoverers came up with these incredibly simple ideas about how the universe behaved. So Newton's Laws of Motion consist of three statements. His law of gravity is a simple law about the interaction of two masses. And all of mechanics and how something moves and how you can use force to get something to accelerate, all came in from the very simple laws that Newton discovered.

And Maxwell, similarly, with the electromagnetic theory which underpins how light behaves, how telephone waves are transmitted, how radios work, what X-rays are, uses four simple equations. So if you look at either of these gems of discovery of the Natural Philosophers then you are completely amazed at how simple and how all-explaining they are.

From the 1850s the industrial revolution happened and from these simple equations, complicated technologies were built up: the technology of the steam engine, the technology of sending telegrams, the whole mobile technology revolution, aeroplanes, cars, and so on, and all these began crowding in on those very simple ideas that the Natural Philosophers had found. So people started thinking, 'Well, this is becoming a very serious thing. There is loads of money being put into all these machines. Electromagnetism can really translate to communication on a mass scale. We really need to be doing something more serious than Natural Philosophy'. So the new industrialists said, 'Let's not call it Natural Philosophy any more, let's call it the Hard Sciences. Let's stop messing around with this part-time type of thinking, let's get serious, we've got money invested in it and its time for a different approach'.

This new approach involved a rigorous application of scientific method, which in 1887 was described by Maxwell as follows:

In all scientific procedure, we begin by marking out a certain region or subject as the field of our investigations. To this we must confine our attention, leaving the rest of the universe out of account till we have completed the investigation in which we are engaged. (Maxwell, p. 2)

What happened when science was made hard and fixed, was that ambiguity immediately came into the content science was discovering. For instance, in introducing the notion of 'fields', the underlying equations could deliver counterintuitive solutions, such as a wave travelling faster than time itself, moving from future to past. This is discussed further below.

In 1905 Einstein showed that an apparent absurdity, the constancy of the speed of light to all observers, led mathematically to a consistent theory in which there was no such thing as a fixed space and a fixed time, but space and time were ambiguous. Depending on your motion, you might see one event happening before another event, which somebody else in a different motion might see happening the other way around. And what you thought as a metre apart, somebody else travelling at a different speed might see as half a metre apart. In introducing matter into the equations in 1915, the equations gave huge insight into the nature of the cosmos.

And then in the 1920s, a mathematics of probability was developed in which there was no such thing as the individual particle with a precisely defined position and momentum. If you sent a particle through two slits then the ambiguity about whether it passed through one slit or the other slit was so real that the possible paths interfered with each other and determined where the particle would end up.

For mathematics to be effective as a tool the theoretician had to be versed in paradox and ambiguity, which were the necessary freedom for any worthwhile insight to be developed. But in order to gain credence from the academic and industrial community, science had to argue it was more than just an allegory. What one had to do was hide the paradoxical approach within a certainty that one could sell.

This relates to the episode with the desk and key related in the introduction. I was a bit worried about having this key, it wasn't very safe, I could lose it or somebody could take it, and so in my childhood creative mind I had this brilliant idea. What if I locked the desk and put the one key through the crack between the drawers into the desk. I was so delighted with this idea, because all the uncertainty about whether the key made the contents safe was gone – the key that disabled those contents ever coming out was in the desk with the contents. So I managed to take this fragile, uncertain thing and by putting it into the content that it was there to protect, I introduced some certainty there.

Something similar happened with the project to develop Hard Science, when they kept on coming across this ambiguity. How to deal with it? So Freud in 1900, for instance, said that it's impossible for the child to be himself as there are many unconscious complexes at work, preventing him being himself. The only hope for him is to see a psychologist and the psychologist will listen to his dreams and bring out the content of who he is and in that way the psychologist will discover what is preventing the individual being himself.

> What a person thinks he remembers of his childhood is not a matter of indifference; hidden behind these residual memories, which he himself does not understand, there are as a rule priceless pieces of evidence about the most significant features of his mental development. (Freud, p. 62)

One can almost feel Freud here pulling the individual (in this case Leonardo da Vinci and his dreams of a vulture) back into the scientific abstract realm, where the masterful creative artist, inventor, Renaissance man, is cut from his infinite aspiration back into finite dimensions of an underlying fantasy at play.

In quantum theory even though the particle doesn't really exist in its own right, we can also predict the limit of how possibility appears at the instant the system is disturbed by measurement.

When I put the key into the desk drawer that was locked I had a sudden moment of understanding what that desk was about. I understood in some definitive way that the desk was there to hold the contents within it. In putting that key in the desk there was that moment of glimpsing what the desk was doing holding the content and how important the key was.

Science is only able to say anything about the world when the way the world makes order is reflected in the order itself. It was thus the trick of psychology or science to stealthily add into that self-generating order, an apparently innocuous action that broke the relation between the mathematical description and the phenomena. It then appeared with this action of measurement, that the causal description was not derived from the phenomena, but actually preceded it as an explanation.

The mind only makes sense of the world by selecting from all of the possible options of the paradoxical world, a simplified conceptual map. Pirsig in *Zen and the Art of Motorcycle Maintenance* defines quality as the process of selection from the paradoxical background to make a conceptual foreground. Our mind is continually filtering the totality of everything we see into a single conceptual picture. For Pirsig this event of quality is pre-rational.

There exists a filter by which the paradoxical world of possibilities appears to us as a simple logical construction of conceptual simplicity. What science does, however, is assume that the map of logical construction precedes all of the paradoxical complexity. Reason becomes lost, according to Pirsig, when philosophy tries to rationalise quality.

And there is a sort of fatal logic in this. By choosing to apply the filter of Hard Science, seeing nothing but the conceptual capability of the mind, all the other possibilities contained in the living phenomena are dissipated. The subject is proverbially killed off by applying a filter that sees only the conceptual construction of reality one already possesses. So the filter one uses to sieve through reality seems to justify one's choice to ignore everything but the prediction made by the theory.

Leonardo became an instance of a theory, or the key became part of the contents of the desk, or the ambiguity of elemental self-reference became measurable in its properties, precisely at the moment at which the outer understood the inner by destroying it.

The interpretations gave understanding at the very point in which the dynamic of inner potential was disabled from acting by being secreted in the containment of a static theory. The scientist walked away whistling as if the theory that had just hidden life was now the reality.

Light

Around 1860 science was becoming organised around its own rational concepts. Darwin's theory of evolution forwarded the idea of life as being its own maker. Science could now finally concentrate on what it could directly prove. Just at this time when science was closing the door on anything one didn't exactly know, there was a theory produced by Maxwell in 1861 about light. Light left the mystery half outside the door and half incorporated into the rationale of science. This was the magical mystery, right in the middle of science, about the nature of light.

In 1861 there were these two explorations going on, one into electricity and one into magnetism. And these weren't obvious phenomena at all. Electricity and magnetism were properties. So something that had a magnetic property exerted a force on other magnetic substances. Something with the property of an electric charge had a characteristic behaviour in relation to other charges. But it was not like matter where you could isolate it, where for instance you could break down a chair and eventually get to the atom and that's what the chair was made of. It was much more like an intrinsic property that resulted in certain behaviours.

And so the early experiments to explore electricity involved amongst other things, putting kites up in thunderstorms and trying to get electricity down to explore its quality. Scientists lost their lives by being electrocuted by the lightning.

> Richmann is more often remembered as the tragic victim of a fire-ball (a globe lightning) which killed him as he watched a thunderstorm. He had studied electrification and electric conduction, discovered electric induction, and invented what he had named the electric indicator – an instrument with which he could compare electric charges in magnitude. (Sharle, p. 47)

Electricity remained a very mysterious phenomenon. Gradually in the nineteenth century it became clear that instead of trying to define what it was, which was very difficult, the way forward was to state how it invoked a relationship. Two unlike charges attract or like charges repel. Faraday, Ampere and others saw that what you needed to describe it was the idea of a field, E – an electric field, or B – a magnetic field, and this field was a pre-impression that you put on reality about how reality would then behave in the vicinity of this charge-source.

The quality of the field determined the kind of behaviour that happened in it. This field wasn't reality itself. It was a tendency as to how reality was going to behave when other charges entered into the field. If you had a positive charge and another negative charge entered into its vicinity, the charge would be attracted along the lines of force in the field of the positive charge. So you could explain this by a field. A field is a pre-impression so that anything in that field will respond to the pre-impression of what it is supposed to do and adjust its behaviour accordingly. Without understanding what the magnet is, the field gives a description of how it affects the behaviour of sensitive objects (iron filings are particularly good at showing the magnetic influence) in its vicinity. The field is indicated by how the lines of iron filings shape themselves around a magnet. So if you put an iron filing in the neighbourhood of a magnet, it will tend to align itself along the field lines, as the magnet has an influence in the space around. This influence is going to determine what happens so the iron filings, sensitive to the magnetic field, form exactly the pattern indicative of the field.

An electric field similarly is a pre-impression, which isn't reality itself, but which is co-produced with a magnetic field.

If a magnet is moved around a wire in the middle, then the pre-impression of a magnetic field causes an electric field in the wire and a current is produced. There is some connection of this pre-impression of a magnetic field to an electric field. So in the dynamo on a bicycle, a magnet goes round from the turning of the wheel and produces a current that runs the light.

Conversely, a coil of electricity will produce a magnetic field, a principle used in the doorbell, where the lever is pulled back by a magnetic field that breaks the current, allowing the lever to fall into place to make the circuit again.

We still don't know exactly what magnetism is or electricity is, but we can define it as a pre-impression or field as to how charges in the vicinity are going to behave.

When Maxwell came along everything was in confusion. What was electricity? What was magnetism? Maxwell came along at a key moment in history when science was about to close the door on anything that wasn't completely material and physical. And Maxwell, to everybody's astonishment, and probably also his own, came up with these four equations describing the mutual creation of electric and magnetic fields. When an electric field changes, it induces a magnetic field, and when a magnetic field changes, it induces an electric field. These equations show how the electric and magnetic fields co-arise. Maxwell hit upon the formula for the arising of the electric and magnetic fields to completely encompass all the phenomena of electricity and magnetism: why you get an electric field when a magnet goes round, why you get a magnetic field in a coil of current. Everybody was completely amazed that these diverse areas of phenomena have four equations that say how the behaviour co-arises from simple relationships. Maxwell gave the essential description which holds the essence of the phenomena of electricity and magnetism.

What Maxwell also realised was that the equations allowed for a wave where E – the electric field when it changes, is going to cause B – a magnetic field; the magnetic field is going to be changing as it arises, so that it is going to produce an electric field; and that this again produces a magnetic field, and this produces an electric field, and this oscillation would present a travelling wave. So starting with a moving electric charge, instigates a travelling wave between electric and magnetic fields, until this motion is received by another moving charge somewhere else as receiver of the wave. In-between transmission and reception, a travelling wave of electric and magnetic fields would move allowing a propagation of this effect along an axis, which is the energy of the wave. This wave had a calculable speed of nearly 300,000 km/sec and Maxwell proposed that light was an example of such a wave. The speed of light was given also at a similar value close to 300,000 km/sec. The speed of light and the speed of the electromagnetic wave were exactly the same.

Hertz carried out experiments applying Maxwell's theory to the production of radio waves. He also showed these artificially stimulated waves had the same speed as light, which conclusively proved both to be electromagnetic wave phenomena. The phenomena of electromagnetic waves (including light) could be generated with a moving charge. This electromagnetic wave could then be received by a moving charge of the right frequency to pick up this wave. Signals could be sent artificially by a transmitting device with a moving charge. This produced a wave. A receiving device a hundred yards away, or in America, if tuned into the right frequency, would receive this signal at the incredibly vast speed of 300,000 km/sec.

The story of how Bell discovered the telephone, while involved in a completely different experiment, gives a flavour of the happenstance with which electromagnetism was investigated and then utilised.

On June 2, 1875 during one of Bell's experiments in which he was assisted by Thomas A. Watson, one of the reeds at the sending end had stuck to its electromagnet by chance (another chance to our score of fortunate discoveries). When Watson tried to get it going again by plucking it (unsuccessfully because the contact point of the reed had welded to its counterpart on the electromagnet), Bell could hear at the receiving end a faint but clear sound not unlike the one given off by a string stretched taut. At that instant the telephone was born. (Sharle, p. 156–57)

All the phenomena of radio waves, x-rays, phone transmissions, in different parts of the wavelength spectrum to the visible signals that our eye receives, are all encapsulated by this one set of equations and the cyclical relation of electric and magnetic fields. But what is it that is relating? We've looked into the heart of the universe and seen this relation of change, which is so fruitful in how we communicate, but what is it that is relating? The electric field is not anything material, it is not *something* that is relating to something else. E is just this field, this pre-impression of what can happen; and B is another pre-impression of what can happen. So E is a potential that predisposes reality to happen in a particular way. B is another type of pre-impression asking reality to behave in a particular way. But there is no material reality here, nothing is being expressed, nothing is carrying this wave. All this wave is carried by is two pre-impressions of how reality could express itself that are simply progressing their way in Maxwell's dance step through existence until they meet a receiver at the other end. There is ambiguity and freedom in the understanding of E and B that resolves only in the context of their expression together.

So light from a star reaches us after its journey of a hundred light years and the reason that it reaches us completely intact, as if we were seeing it with the clarity of a nearby object, is that it is not actually in physical reality on its journey. Light is held in the translation of these potential impressions of how reality is going to behave. Only when light meets our eye does it become again expressed as something tangible and enter into physical reality. This was Einstein's 'aha!' moment; that light is in reality – it arises from something physical happening, it is received by something physical happening. But in between transmission and reception, the relation between fields isn't itself in physical reality. It is a predisposition to reality.

Light has a completely special role with regard to physical reality – half in reality, half out of the door of the physical. On one hand, we have this explanation that allows us to create phones with a tiny transmitter sending a signal, to communicate to anyone in the world. There's nothing we have been able to grasp and control more than electromagnetism. On the other hand, the parts that are involved in making this work are not material – we know how it relates but we don't know what it is.

Light is the messenger that is half in the world and half out of it, a mythical creation, half in physics and half out of it. Einstein used light to see beyond the absolute framework of

Newton, absolute space and absolute time as the stage where things happened. Einstein realised that light had this property of being in the universe of separation and yet outside it. Light is a reference to which things can relate outside of material reality. The hypothesis of an absolute framework to physics was refuted by light.

When physics closed the door, there was this one key to where it had come from, its mystical origin, of things existing beyond what we know, half left out of the door, a clue that we can follow to return to something that is more than what we can explain.

Two ways

Much like Newton has done for motion in his *Principia Mathematica*, Maxwell found the generative dynamic of nature, deep within a mathematical inquiry.

It is important to understand what this achievement signifies. Mathematics is able to identify a generative dynamic of potential that precedes the physical consequences that follow from it. Purely from mathematical elegance alone, Maxwell, as Newton before him, found the very character of a field of study within the dynamic of conceptual interactions. For Newton, the kernel had been the laws of motion. For Maxwell, it was four equations that together gave the full generative description of electric and magnetic field co-arising.

The mathematics managed to grasp the essence of the conceptual play that was able to explain all observed behaviour from nature as causal consequence.

One of the huge surprises of Maxwell's mathematics was that it demonstrated the existence of electromagnetic waves that were able to propagate a message without any physical carrier. The generative capability of the electric and magnetic fields simply swapped around their potentials as a matter-free cycle of transmission. (It is hard to remember that before this, the horse was the fastest way of delivering a message).

Yet, just as we found with the two possible physical interpretations of Newton and Leibniz, there were in Maxwell's equations two solutions of how this generative possibility would be expressed.

In the first solution, the electromagnetic wave moves moment by moment in a causal progression matching step for step the physical reality of any material wave. This solution is known as a retarded wave, or *Time in Light*. The wave moves with a particular (retarded) speed around 300,000 km/sec.

But a second solution exists, just as Leibniz understood change through the monad. In this second solution, the ambiguity of definition of the field is left open as to its meaning until the whole is resolved in completion of the journey. In this second solution, the wave appears to travel *ahead of time*, to meet at the limit of the exercise of its inference, meaning acting back to confirm the propositional elements. This is known as an advanced wave.

As Huw Price writes:

> Maxwell's theory of electromagnetism, developed in the mid-nineteenth century, is easily seen to admit two kinds of mathematical solutions for the equations describing radiation of energy in the electromagnetic field.
>
> One sort of solution, called the *retarded* solution, seems to correspond to what we actually observe in nature, which is outgoing concentric waves. The other case, the so-called *advanced* solution, describes the temporal inverse phenomenon – incoming concentric waves – which seem never to be found in nature. Thus the puzzle of temporal asymmetry here takes a particularly sharp form. Maxwell's theory clearly permits both kinds of solution, but nature appears to choose only one. (Price, p. 50)

Einstein accounted for the absence of ever seeing advanced waves of light by remarking on the impossibility of these future originating waves ever to arrive in coherence at a present observer. The future could not act locally on the past or it would change the past in an inconsistent way.

Wheeler and Feynman

Wheeler and Feynman made a new sense of the advanced and retarded waves in their paper *Interaction with the Absorber as the Mechanism of Radiation* (Wheeler 1945). The timing of this paper is noteworthy in that Wheeler was at the same time working at Los Alamos on the Atomic Bomb project. Feynman was a prodigy, his star student destined for the Nobel Prize.

Feynman in his Nobel Lecture acceptance speech describes their work as follows:

> So, one day, when I was working for Professor Wheeler I calculated the following. Suppose I have two charges – I shake the first charge, which I think of as a source and this makes the second one shake, but the second one shaking produces an effect back on the source. And so, I calculated how much that effect back on the first charge was, hoping it might add up the force of radiation resistance. It didn't come out right, of course, but I went to Professor Wheeler and told him my ideas.

> But, as I was stupid, so was Professor Wheeler that much more clever. First, he said, let us suppose that the return action by the charges in the absorber reaches the source by advanced waves as well as by the ordinary retarded waves of reflected light; so that the law of interaction acts backward in time, as well as forward in time. I was enough of a physicist at that time not to say, 'Oh, no, how could that be?' For today all physicists know from studying Einstein and Bohr, that sometimes an idea which looks completely paradoxical at first, if analyzed to completion in all detail and in experimental situations, may, in fact, not be paradoxical. So, it did not bother me any more than it bothered Professor Wheeler to use advanced waves for the back reaction – a solution of Maxwell's equations, which previously had not been physically used.

I found that you get the right answer if you use half-advanced and half-retarded as the field generated by each charge. That is, one is to use the solution of Maxwell's equation which is symmetrical in time. (Feynman 1965b)

Wheeler's and Feynman's understanding of advanced waves involved a transaction between emitter and absorber in which a half-wave went forward in time, was received by the emitter and a half-wave returned. The sum total of these constructively interfering two waves was then of a whole wave heading from emitter to absorber forward in time, causally consistent.

This reasserted the symmetry of future and past without affecting causality.

For Wheeler and Feynman, each event consists of one advanced half-wave – a darkness that is travelled ahead of time from the absorber, and a retarded half-wave – a causal path from the emitter event. The condition of physics amounts to those solutions where the two half-waves combine synchronously into a viable whole wave. The interchanging half-waves (retarded and advanced) combine into a whole understanding of light's passage through the intervening space and time.

Wheeler and Feynman made sense of advanced-retarded waves as: (1) a retarded half-wave is sent out from an emitter; and (2) an advanced half-wave is sent by the absorber to an earlier point of time. By imposing the condition that the two half-waves then cohere, the effect is the same as if the receiver had sent a whole wave out to the absorber, as is observed!

For instance light travels 100 years from a distant star to reach us. The Wheeler-Feynman interpretation would be that there is a handshake from us in 2015 to the origin of the light when it set out in 1915 that ordered the photon to make its journey back from then to now. There is reciprocity between seeing and the star seen, that transcends the idea of information arriving to us over a distance of 100 light years. Seeing allows us to elicit in an advanced half-wave question, the information the star needs to send to fulfil our inquiry with sight.

The idea of Wheeler Feynman combines inner potential with outer actualisation. The act of seeing requires both a retarded-half-wave of the outer information of the star's existence; but also the advanced-wave with the question of concern to inner potential included in its whole demand upon reality. Over the inner question and the outer answer, a whole forward-moving wave is observed to proceed from outer source to inner witness.

Seeing and sight are two aspects of the same process that combine in the full wave of light. Seeing stimulates the observed to respond with a time-forward response of being seen.

Perception and meaning

The conceptual reality of our mind gives perfect example of Wheeler-Feynman's physics. From the multitude of inputs coming through our senses and internal processing, a filter must operate in a process that ends with a conceptual picture. Not only are we not aware of this process but the end result appears to us to have its origin in the world we see around us. In arriving at a conceptual picture of reality, a process of seeking is resolved by an illumination, which precedes the content of what we then experience ourselves as seeing.

The freedom of association of our conceptual interpretations are such that they allow for the whole meaning of the undifferentiated reality to settle coherently into our perception. Modern neurology separates the mechanism of our seeing, with the processing of the brain to arrive at meaning. But the meaning and the seeing are evidence of the combination of advanced and retarded waves that we have encountered with Wheeler and Feynman. The net of concepts have a twofold action:

1. The concepts are suggested in their separate ambiguity that a unity may signify their individual explorations together as meaning. The whole acts back to fulfil the explorations of the ventures of the parts into the unknown with a retrospective completion.

2. The concepts draw authoritatively from undifferentiated reality a particular interpretation, accessible to concrete analysis, of what happened. The parts act in retarded, causal inference to motivate together a finite sense.

The encounter with the unbounded nature of reality can only be translated into our finite perception, if it can be fitted coherently into both the unity of abstract causes of the past as well as the surrendering of finite signs into the significance of the future.

As Bortoft showed in Chapter 1, our seeing is able to switch between two alternative views, that of the unity of the whole phenomena and the details of the particular trajectory. The combination of these two perspectives occurs naturally in our seeing. The very quality of seeing reveals a unified vision to a complex web of particulars. Instead of interrupting our vision and interpreting it, seeing by itself works out a path that is inclusive of both unity and particularity.

The mix of meaning and understanding is innate to sight. But this combination of perception and meaning is not simply about thought or mind. Also in nature, as confirmed by the solution to Maxwell's equation, this foreseeing of the conceptual basis on which time extends, is everywhere in play. In the diagram below we imagine the advanced wave from the future and the retarded wave of the past coming together at a point of agreement on the present. The future is tapped for its version of events, to the extent that the past, as an incomplete riddle, leans to the future for resolution. This is shown in *Figure 9*.

Instead of seeing existence as something solidly present, we understand that an identity is rooted in the past to the extent of the reach of its becoming into the future.

The Feynman-Wheeler template shows how two differing waves, the future net of meaning and the past actuality of an event, move towards each other to provide a consistent pattern of happening.

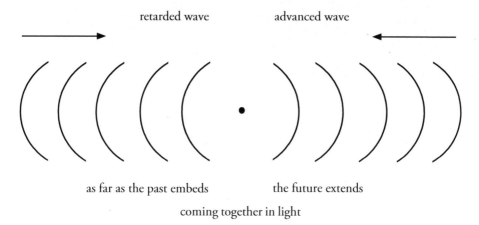

Figure 9. Retarded and advanced waves.

Light tells us about time equally to the extent that time tells us about light. Whatever we glean from the past about where we have been, in a description of a temporal event, is balanced by an illumination from the future that tells us the meaning of time in the events travelled to take us there. The address of whole nature exactly balances the extent to which we are able to look into the past to understand where we have been.

Advanced quantum theory

John Cramer in his *Transactional Interpretation* applies an advanced-retarded dynamic to quantum theory.

The way one understands quantum theory is through a possibility left open in Schrödinger's wave function as to the potential nature of reality. The wave propagates itself just as the electromagnetic wave would do.

Cramer in his Transaction theory, as Feynman Wheeler before, understands that every quantum event is a mixture of a novel meaning being integrated into a habit of the past. This is shown in *Figure 10*.

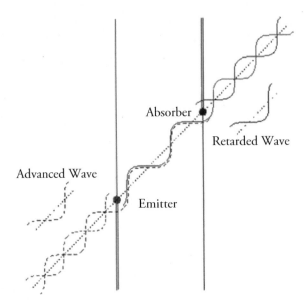

Figure 10. A plane-wave transaction between emitter and absorber (from Cramer).

The transaction interpretation then sees time as the synchronisation of past histories with a future unity able to establish a common foundation to event in light. The fact that the wave function incorporates a future action of unification was further elaborated by Bohm and Hiley in their work on the quantum potential field.

Quantum potential field

David Bohm and Basil Hiley described the advanced action of the quantum wave through the quantum potential field. The quantum potential field reverses the usual understanding of order as something causally unfolding from the past. Order arises through a freedom of the future able to allow in its midst for a whole form to be accommodated. The quantum potential field is simply the order of accommodation in which an event of meaning is coherently distributed over the available

Slit A   Slit B

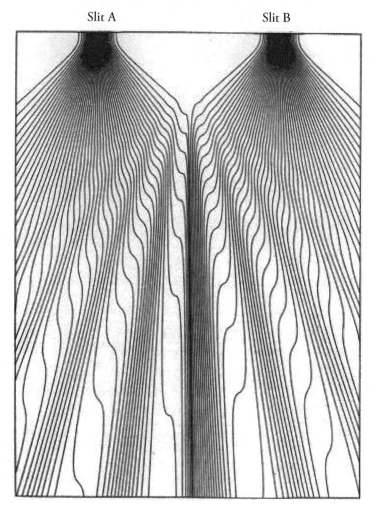

Figure 11. Quantum potential: trajectories of a particle in a double-slit experiment. (Bohm & Hiley, p. 53)

possibilities. The field channels the occurrence into a future order whose relatedness absorbs the significance of the whole, undifferentiated impulse. This is shown for the double-slit experiment in *Figure 11*.

The quantum potential field arranges the freedoms to accom-modate the act of the particle travelling through the slits, so that the ambiguity in the individual responses overall accommodates the whole undifferentiated occurrence. Basil Hiley explained this:

> We found that the potential was totally different from any classical potential that we know. It has no external source in the sense that the electric field has its source in a distribution of charges. It does not act mechanically on the system. In this sense it cannot be thought to act like an efficient cause. It is more like a formative cause that shapes the development of the process. Indeed as we explored its properties in many different physical systems it reminded me very much of the morphogenetic field generated in biological systems.
>
> The information field is shaped by the environment in a way that is very similar to the way the development of a plant is shaped by its environment. Thus we can think of the information as active from within giving shape to the whole process and this shape depends on the environment in key ways. In other words the meaning in the wave is expressed through the form that develops. (Hiley 2004)

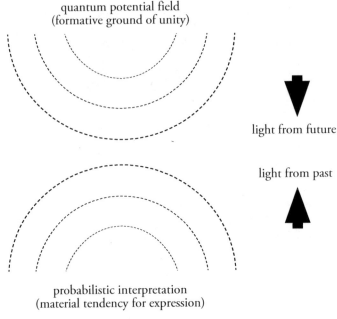

Figure 12. *Quantum potential field.*

The unity is no longer a realisation of the mathematics and measurement. Unity is now actually physically located in the way the future is able to exercise a coherent meaning upon the contributing elements. This is shown in *Figure 12*.

Wholeness is always in the allowance of the future to speak back to all the individual lines of inquiry.

Just as Bohr succeeds in capturing the possibility aspect of information into a causal (at least at statistical level) quantum law, so Bohm suspends causality for the entrance of an informational field that steers free flowing potential. The contrast in the two perspectives is summed up in *Table 1*.

Bohr's concepts	Bohm's meaning
hides phenomena in an elusive world of statistical average	celebrates form in wholeness
causality breaks multiply into infinite possibilities only recovered by a statistical averaging	causality breaks singularly and dramatically to restore causality as after-influence
aligns a global nature to permeate the weird world of local phenomena	local accommodation of freedoms allow the potential for a unitary resolution in light
alerts us to the stability of the atomic founded substance over all inducement to change	triumphs momentarily over history to cohere potential into new form

Table 1: Comparison of Bohr's conceptual model with Bohm's informational perspective.

Current experiments by Basil Hiley at UCL follow on from work by a Canadian team (Kocsis *et al.*, 2011) using a technique called 'weak measurement' to 'see' *both* the position and momentum of a photon as it passes through the double slits. The conclusion of such experiments is that separation and wholeness are both active in the phenomena simultaneously. If shown to be true it will give conclusive verification to the quantum potential field.

Another role for matter

The understanding of Wheeler-Feynman has also been applied to relativity as a basis for an explanation of dark matter. Relativity is given natural extension in the additional paradox of retarded-advanced waves.

Dark matter, together with dark energy, was first hypothesised in the 1930s to account for the mismatch between the gravitational behaviour of the universe and the amount of matter actually visibly present. Modern observations have since estimated the amount of dark matter and dark energy together to account for about 96 percent of the universe's total matter! The matter and energy are called dark because they do not interact with light. In other words the matter is invisible to any telescope or radio receiver or any act of seeing. Dark matter passes through other matter. However the matter is gravitationally effective and this can be inferred from its distortion of light or other matter in the surrounding universe. Experiments to test whether the dark matter and energy may be responding to another type of force, other than light, have so far all been negative.

Another approach into this field is known as the bi-metric ('double solution') approach. The order of dark matter and energy may be compared to the notes of a score of music. Individually the notes have no established reality. But the association of their various freedoms and ambiguity of harmony allow that some whole, undifferentiated identity may be held in the interactions. Dark matter and dark energy are similarly the backdrop of possibility that absorb into them a certain freedom that allows some whole symphonic sense of the universe to sound. Ripalda in his paper *Time reversal and negative energies in general relativity* begins by referring to the Feynman-Wheeler work:

The equations of classical physics show no preferred direction in time. The possibility of particles propagating backwards in time is inherent to all relativistic theories, quantum or not. Wheeler and Feynman found that a consistent description of radiative reaction in classical physics requires the usage of past-pointing 'advanced action' on an equal footing to future pointing 'retarded action'. (Ripalda, p. 1)

Instead of starting with the concepts, in this case space, time and matter and seeing how they behave in causal progression through their laws of interaction, one can argue the other way around. A tension of potentials waits for definition from the whole movement of the future. In this case, the concepts are described backwards from the unified event of meaning into the separate conceptual identities.

The geometry of space-time has two branches. Time reversed matter moves on a different branch than future pointing matter. On a large scale, equal amounts of future-pointing and past-pointing matter should be expected. (Ripalda, p. 4)

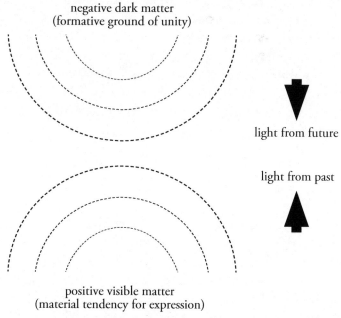

Figure 13. The bi-metric account of dark matter.

Matter is given a negative sign in the way it waits for future meaning to accord its potential a definition within other aspirations. The negative matter has a repulsive gravitational influence on visible matter. This is shown diagrammatically in *Figure 13*.

Dark matter becomes a state of matter that responds to the illumination of a future meaning. This negative matter disperses itself so that a story of the universe can call it into being most appropriately. Light is met through the adventure of chance, to become the vehicle of coherence. Some part of all existence, planets, earth, life, human spirit, is this receptivity to respond to light in the venture of experience.

Dark matter is the unformed relation to the universal test of light the future brings. The proportion of dark matter compared to visible matter gives an estimate of how far the universe is working towards the inclusion of freedoms into a cradle of new form and how much space is established giving home to existence already formed. The proportion of space existing in the holding of negative matter of potential forward into light is of the order of 25 times greater than the space housing positive matter, fully formed!

This process is illustrated in moments of realisation where our actions seem to fulfil a quality beyond what we historically are. Ambiguous aspects of the whole shape themselves together with other aspects, into a resolution that holds a meaning beyond the parts. What is lived is then a totality of all that time ever accumulated, spinning itself one thread then another into the inevitability of a confrontation of light with dark.

Event ahead

Having returned from the African adventures of Chapter 2, I found myself in an open-plan computer office in Holland asking how the past can be held in such a routine description.

After several months of such routine, I explored the outside world and ended up from my base in Rotterdam Holland, hitching and reaching a place near the Alps. Such was the pull of the mountains, I left my colleague with whom I was supposed to be going to Berlin and headed south instead. I was left, as a result of this, walking through a rain-swept Swiss landscape wondering at my decision to view clouds from this new perspective.

My attempt to live in paradox on my return to Europe now seemed to involve a much larger choice than just myself.

The weather closes, the greyness oppresses and the dullness resists showing any direction. The clouds curtain off the view one senses behind. Then at a certain moment a coherence enters into the process of thought. It is as if the thoughts seed the greyness with an expectation of lightening. For along the very passage of a new ordering, the clouds themselves begin to open out. The separation of I in isolation shut out from the world, turns into a sudden inclusion. The world enters into a process of showing I along the same lines that nature uses to grow. Now a different I is shown to light, one that has a place in the nature of being itself.

The I whose boundaries dissolved in the mist, is gifted a gentle-edged quality, a discovery made in the journey of changing the aspect of the world to reveal its future inclination. A unity from the future calls to existence, through the veil of a dying and rebirth. The first showing of light through the mist has a direction, pointing to an inhabitancy of the present that is open to show through it an undifferentiated whole statement of reality. The ray of illumination points through the world to a placing signified beyond any current theme. I is pointed to in the circle of light with a significance beyond itself. The very trial of I to focus itself through the mist, clarifies the future to show a universal character. A meaning is made transparent through the very act of working through the dismissal of the grey oppression.

The lightness, that enters, inhabits a world whose effort is distinguished from its result. The working of nature is given over to another end than itself. The light is a miraculous bonus, where everything collects at a focus where the nothing gathers into being.

The step between cause and effect is completely passed over, as the inseparability of the mist dissolves into the end of the pooling of light. Across this emptiness, the future has invited I to act in a sudden discovery of being. (Franses 1982)

Premonition

This sense of knowing an identity as if spoken through future event, was enforced when travelling anew to the mountains.

Hitching from Rotterdam, I arrived into Aachen at 1.00 a.m., just inside Germany from the Dutch border. I chatted hopefully in a bar, where no one invited me to stay the night. From 3.00 a.m. I walked the streets until the sky lightened with dawn at around 6.00 a.m.

When dawn came I struggled to the road and relied on various charitable drivers to take me slowly into Germany. I then met two dynamic young lady hitchers, in whose company I sped right up to the border with Austria.

At the border with high mountains before us, the girls found a ride to Italy, their destination. Meanwhile the driver of the coach that had taken us to the border offered to take me as its only passenger on to the village high up in the Alps to which he was heading.

As night fell, the darkness of the unknown, the lights shining on foreign names and the twisting, windy roads taking us higher, concentrated my earlier feeling. Palpably, as we drove outward through the dark, I journeyed inward.

The sharply contoured world, calm in its stillness, brought out the hills of my own existence inside. The hills of my own existence, of forgotten people, of forgotten places, of forgotten happinesses and sadnesses, of forgotten laughing and attaining.

The hills of my existence sat inside, while the coach drove through the dark.

My being, turning with the coach into the mystery of the mountains, received a distant foundation to myself in the purity of an action.

The coach drove through a town of hotels, then started climbing up hairpin bends, the beam of its headlights swinging across green forest. The lights of the town were left below.

The mountains were around me. I knew them to be there. I had moved beyond them, I could look down inside me and see them there.
I was left above them, looking, from one window to the opposite window, into the darkness. And there amidst my excitement, inside me, grew the uncertainty of a step.

A week in the mountains. Something must happen. The stage had been cleared. An action, whose nature I could not fully see I knew to await me. The shell of my life had opened to a glimpse of its purest centre. (Franses 1982)

Identity could allow the unity of meaning to specify in the world of concepts the nature of my relation to future unfolding.

Renewal

The nature of paradox is open, oppositions are considered and respected as the basis of experience. But the computer programming office prides itself on knowing the one way that any dynamic system can be reduced to its logical parts. The paradoxical nature of reality is denied as an ignorance of the past, before reason and the scientific method had properly understood the world. So what has happened to paradox? In some way the paradoxical nature of the world, expunged from the past, now resides purely in the future.

The mountains communicated to me a meaning whose significance was unclear, but which illuminated a path that was open for me to tread. What is more, the meaning communicated by the future allowed a synchronicity with others contributing to the same meaning. Time was foreseen in the premonition of a meaning that was not yet lived. The meaning allowed the dimension of happening, a reference as an ordering of time.

> Tree do not be sad, your leaves are falling
> Even in your summer glory,
> You knew your calling
> Risen from the dust, there you return
> Defined by space and time

If we look closely at the world, then the initial state of existence is one of ambiguity. The world requires a medium such as light to form itself. But the way light takes on this role is two-sided. Light provides a past-directed aid to observation, accommodating different individual perspectives of space and time. Light may also be future-originating in the converging of aspirations. As in a well-constructed poem, light crosses the plain of ambiguity in a common arrival at coherent meaning, fulfilling the passage.

There are, through this double nature of light, two ways to construct identity. In one we base all that happens on a conceptual characterisation that fulfils its function in any social situation. In the other, light realises our identity through a path we do not yet know. The physics we have understood as a causal development from the past extends to live an undifferentiated meaning in the ambiguous freedom of the future.

Wheeler, Feynman:

←——————————— past/future ——————————→

The Dice of Renewal

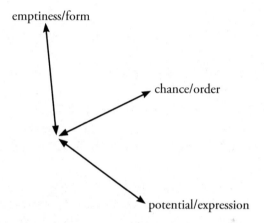

Chapter 4
Chance faces order: the fall

Barfield compares Greek and Oriental foundations to existence:

> The Western outlook is based essentially on that turning of man's *attention* to the phenomena. This is sharply contrasted with the oriental impulse to *refrain* from the phenomena, to remain, as it were, in the bosom of the Eternal, to disregard as irrelevant to man's true being, all *that*, in his experience, which is based on 'the contact of the senses'.

> It is clear that the way of the West lies, not back but forward; not in withdrawal from contacts of the senses, but in their transformation and redemption. (Barfield, p. 173)

Taking as starting point the literal truth of separation of subject and object, the western outlook introduces an alienating sense of the absence of participation. This leads to the apocalypse of all things gathering together a vision of nothingness in their fragmentation. In this void, a way forward is found in the apparently insignificant part played by chance. Chance becomes the key that translates nothing into existence.

Zero

The decimal numbers, in particular the *zero*, had come to the West in surprisingly recent history. They had been used for thousands of years before that in the East. Mathematicians such as Aryabhatta and Brahmagupta first worked with mathematical symbols and treated the *zero* as a number in its own right.

> When *Sūnya* is added to a number or subtracted from a number, the number remains unchanged and a number multiplied by *Sūnya* becomes *Sūnya*. (Brahmagupta AD 628, quoted in Barrow, p. 39)

The designation *Sūnya* for zero suggests the pivotal usage of the term when comparing its other meanings. Alongside emptiness or absence, *Sūnya* stood for:

> ... space, the firmament, the celestial vault, the atmosphere and ether, as well as nothing, the quantity not to be taken into account, the insignificant. (Ifrah, p. 36)

Zero was also the space developing the potential into existence. Space is the ground of becoming, without existence for itself but holding the emerging of the possible. The *zero* was the denying of individual existence that could still find its way to being through the collective.

Zero thus had the connotation of that which reveals the one. Nothing is truly ours. Part of our existence is the ability to fall totally out of existence. The shadow that finite existence throws over itself, is exactly what allows us to know the Divine in whatever form. *Zero* is acknowledged as a meaningful, essential quality that brings to the finite, the humility to know the Infinite.

The symbol zero (known as *sifr*) came to the Arab world through Muhammed 'Abu Jafar' ibn Musâ al-Khowârizmi in ninth century Persia.

The Italian Leonardo Fibonacci then discovered the decimal system of counting when travelling with his father who was representing Pisan traders. The book *Liber Abaci* begins:

> While I was still a child, when I had been introduced to the art of the Indians' nine symbols through remarkable teaching, knowledge of the art very soon pleased me above all else and I came to understand it. (Fibonacci 1202)

There was however a surprising lethargy for doing anything with
the number system Fibonacci introduced into Europe. It was only
with the Renaissance three hundred years later that the decimal
number system took over from the cumbersome Roman numerals,
even for trade. Aristotle, who knew of the Eastern way of counting,
made an argument taken up later by the Church that denounced
the zero as a mere idea. Only around the fifteenth century or so did
it become fully accepted in the West.

One

In Greek time, number was related to the *one*. Aristotle says:

> Every quantity is recognised as quantity through the one, and
> that by which quantities are primarily known [as quantities] is the
> one itself; therefore the one is the source of number as number.
> (Aristotle, in Klein, p. 53)

All concept and number are an abstract expression that follows
from the essential unity of the world.

> The one is the source of number as that which gives each number
> its character as a 'number of …,' thereby rendering it a 'number'.
> (Klein, p. 53)

This unity innate to the world is in its essence expressed as
proportion, harmony or identity. All number follows as corollary to
the one that is in the world.

> The whole realm of number is a progress from the unit to the infinite
> by means of the excess of one unit [of each successive number over the
> preceding]. (Domninus AD 413, 5f, in Klein, p. 52)

Thus number for the Greeks described a secondary multiplicity
to the essential uniqueness of things that existed in the world.

The proportions, ratios and harmonies had a practical goal in realising the aesthetic, giving form to the natural order.

The Greek *one* was something totally in the world. The doctrine of the *one* could be taken up into the understanding of Christianity, where the *one* was transcendent over existence. In putting emphasis on existence through the transformed *one*, the *zero* could no longer be the gateway to the Divine as it was in the East. The basis of the *one* that developed in modern science in quantum theory, collided head on with a doctrine that seemed closer in spirit to the Eastern idea of emptiness.

Complementarity

In the double-slit experiment described in Chapter 1, when one sees reality through the concept of a wave, one witnesses a wave-like event. Change the apparatus to catch the phenomena in its actions as a particle and one witnesses particle type behaviour of something travelling a particular path. One could make complementary lists of different pairs of concepts that could reveal different aspects of the situation. One could have a position pair of glasses or a momentum pair of glasses, but not both. As Pauli states:

> One can look at the world either with the position-eye or one can look at it with the momentum-eye, but if you will simultaneously open both eyes, you get lost. (Pauli 1926)

Something bizarre has happened here. In insisting on the only reality being through the concepts, the understanding needs a further mental construct to explain why these concepts no longer provide a whole view of reality. The concept of position or momentum that was originally an invention of physics to describe objective reality, now gives one only half of the picture.

Bohr introduces a further mental construct, picked up from nowhere, the principle of *complementarity* to explain how the existing mental constructs no longer establish a whole view of reality:

The fundamental postulate of the indivisibility of the quantum of action forces us to adopt a new mode of description designated as *complementarity* in the sense that any given application of classical concepts precludes the simultaneous use of other classical concepts which in a different connection are equally necessary for the elucidation of the phenomena. (Bohr 1929, p. 10)

Greenstein and Zajonc sum this up clearly as follows:

Imagine writing down the set of all true statements about a given physical situation. According to classical ideas, these statements could be written down in one big list; this list would constitute a complete description of the situation. But according to Bohr, we are required to write them on two half-lists. Furthermore, we must enter them in such a manner that to each entry on one half-list, there corresponds an entry on the other. The particle-nature of the electron goes on one half-list, its wave-nature on the other; the position of a particle on one half-list, its momentum on the other; knowledge of the path a quantum took on one, the possibility of interference on the other. Bohr's principle of complementarity insists that we can choose either one half-list or the other – but never both. (Greenstein & Zajonc, p. 95)

This was the final irony. In order to serve the *one* and deny the *zero,* Bohr had now divided logic into two stories about existence! What we were seeing here was a fundamental negation of the premise of the hypothesis of the supremacy of the *one* as a basis for mathematics. In the neat way of a mathematical proof, we had shown that assuming the *one* to be absolute, there had to be two stories of existence, denying the original supposition.

Joining

In the East, *zero* or nothing was the natural reflection of a purely spiritual attainment, placing the wellspring of existence outside the content-filled material occupation of things. Nothing was simply the point of natural balance between spirituality and form. This was a reflection of the understanding that emptiness, from which form came, was dynamically accessible to the spirit.

> Form is emptiness, emptiness is form
> Emptiness is not separate from form, form is not separate from emptiness
> Whatever is form is emptiness, whatever is emptiness is form. ('Heart Sutra' from Prajnaparamita-Sutras)

In the consideration of Western science, *zero* is that mediating number between abstraction and expression that is able to deliver a virtual thesis about existence. The decimal system was gifted to the human intellect to know the world through reason alone. The power of nothing was still there, but its home was now the reasoning that was able to apply the concept as structure to the world of nameable things.

In the original Eastern understanding as far as this is known, the *zero* was a spiritual experience of travelling through a state of nothing to illumine a reality of identity behind. There were here two equally valid applications of number to the interpretation of a physical system. One was material, grasped intellectually; the other spiritual, about the nature of being.

The result of the import of numbers from the East was that *zero* was then considered part of a number system, an element that behaved according to the rules of addition and multiplication consistently and thus could be considered as equally real as any other quantity that was actually there.

Zero was simply the absence of any number or when included as a digit in a larger number, an empty designation for a particular power of 10. There was nothing mysterious or creative in the *zero*.

The contortion of western mathematics is that it tries to impose the *zero-one* code of arithmetic as a unique way of making sense of what is there. Rather than allowing the dynamic of *zero* and *one* to be a many-fingered weave of the fabric of structure between darkness and light.

Communicating emptiness

The problem of the use of concepts lay within the place of *zero* within number itself. While the East understood that emptiness and form were connected and hence the concepts of physics would have to hold the dance between unity and separation, for the west, number as abstraction has set itself up as a complete window on existence.

The notion of *zero* developed in the East had a profound spiritual quality to it. It was the fulcrum on which spiritual understanding could encounter the finite by surrendering to the infinite of nothing.

Bohr and Heisenberg travelled to India to meet spiritual leaders and felt vindicated in their philosophical premises by the spiritual tradition of India. Maybe they did not realise that their mathematics had its origin in just this tradition in the first place.

> In 1929 Heisenberg spent some time in India as the guest of the celebrated Indian poet Rabindranath Tagore, with whom he had long conversations about science and Indian philosophy. This introduction to Indian thought brought Heisenberg great comfort, he told me. He began to see that the recognition of relativity, interconnectedness, and impermanence as fundamental aspects of physical reality, which had been so difficult for himself and his fellow physicists, was the very basis of the Indian spiritual traditions. (Capra 1982, p. 217–18)

> Heisenberg told me that these talks had helped him a lot with his work in physics, because they showed him that all these new ideas in quantum physics were in fact not all that crazy. He

realized there was, in fact, a whole culture that subscribed to very similar ideas. Heisenberg said that this was a great help for him. (Capra 1989, p. 43)

When Bohr and Heisenberg each travelled to the East, they recognised some familiarity to their formulation of quantum theory. As HH the Dalai Lama writes in *The Universe in a Single Atom*:

> According to the theory of emptiness, any belief in an objective reality grounded in the assumption of intrinsic, independent existence is simply untenable.

> All things and events, whether 'material', mental or even abstract concepts like time, are devoid of objective, independent existence [...] Things and events are 'empty' in that they can never possess any immutable essence, intrinsic reality or absolute 'being' that affords independence. (HH the Dalai Lama)

This rings true with quantum theory, where reality exists only in so far as it is observed. There are no fundamental entities, such as the particle, or even properties as position or momentum by themselves.

> The actual teachings on emptiness imply an infinitely open space that allows for anything to appear, change, disappear, and reappear. The basic meaning of emptiness, in other words, is openness, or potential. At the basic level of our being, we are 'empty' of definable characteristics. (Tsoknyi Rinpoche, p. 120–21)

How clearly this paragraph changes the null connotation of emptiness with the creative sounding words of openness and potential. Potential exists independently of how we marshal it into being. To be open to the world means also that we do not prejudge anything about it, but allow the world to register its own meaning upon us.

However much Bohr and Heisenberg could intellectually grasp the meaning of emptiness as practised in the East, they were committed to providing a conceptual statement of an objective reality, as required for a western scientific establishment. Emptiness was thus included in a token ring of philosophical statement whose purpose was to safeguard the concepts science had built up in the tradition of Aristotle's *one*.

Colour

The bridge between *zero* and *one* is further provided by Goethe's work in the relation of dark and light.

Johann Wolfgang von Goethe (1749–1832), as well as being an outstanding poet and playwright, also developed a methodology about how one could constructively engage with the world of appearance. His participatory approach differs from the immediate prescription of things one can read from a common science book in requiring a commitment of the whole being. Goethe invites us to experience light where it meets darkness.

As Rudolf Steiner said:

> For Goethe darkness is not the completely powerless absence of light. It is something active. It confronts the light and enters with it into a mutual interaction. Modern natural science sees darkness as a complete nothingness. According to this view, the light which streams into a dark space has no resistance from the darkness to overcome.

> Goethe pictures to himself that light and darkness relate to each other like the north and south pole of a magnet. The darkness can weaken the light in its working power. Conversely, the light can limit the energy of the darkness. In both cases colour arises. (Steiner, 1897)

Goethe's theory of colour was in contrast to Newton's theory. Newton felt the colours were in the light, separated out by the prisms into their constituent elements. In this case, light is merely an abstract player and there is no inner story about its origin.

... they [Newton and others] maintained that *shade is a part of light.* It sounds absurd when I express it; but so it is: for they said that *colours,* which are shadow and the result of shade, *are light itself.* (Goethe as quoted in Eckerman)

For Goethe, the colours were the richness at the boundary of dark and light. To understand colour as a part of light as in Newton's explanation, was to miss the rich quality of the in-between space of darkness and light. Colour was about the dynamic of rich event between emptiness and existence, as they played through each other.

But what had happened with science was that colour had been taken from this dynamic in-between realm and put back into the world of explanation as just an illusory by-product of properties such as wavelength. For Goethe, this explanation hid the true nature of colour, which he saw as fundamental to form.

Something similar happens with chance.

Chance

Chance is something that comes out of emptiness. Chance is not yet based on existence, even though that might be the outcome. Chance is something that happens, whose significance is provided by its relation to other events. Chance events often begin by losing oneself, having strayed from the normal path, to be redirected by an encounter with a whole quality to existence. Chance events thus stand between nothing and the unity of existence.

What Bohr and Heisenberg recognise is that quantum theory is about chance. One understands the equations by talking about the probability with which something chances to happen. But also they have the responsibility of holding up a theory that the scientific establishment shall accept. How can they just talk of chance?

The skill in the Copenhagen interpretation is to introduce subtly through the back door, the act of measurement as inseparable from the results the theory produces. Measurement confines the chances that can occur exactly to those directly quantitative outcomes that can be translated into mathematics. The theory is still about chance but now the range of events is limited to those that return a certain numerical answer.

The one thing Bohr was good at was completely disguising the meaning of what he was saying. In his papers it is difficult to pin down what he means when he gives an explanation. It is always just beyond what you think you've understood. He either doesn't completely explain what he means by measurement or by *complementarity*, or he says it in a way in which, when you've read the sentence, you can't quite get at what he was saying.

The last broadside attack on Bohr's theory from Einstein was the Einstein, Podolsky and Rosen (EPR) paper in 1935 which broached the subject of entanglement. Two particles *A, B,* initially together, fly off in different directions and when a distance apart, one property of *A* is measured; as measurement fixes a probability into some actual manifestation of a value and since the sum of *A* and *B* must have a conserved quantity as they began life together, knowing *A* instantly influences the value that registers of *B*. The probability spectrum at B is resolved, *without the necessity for any measurement.* This experiment required Bohr to dilute the notion of local measurement as the glue of reality that holds the mathematical construction together.

John Bell, the physicist who gave mathematical rigour to the consequence that entanglement has for reality, recalled:

Bohr reformulated the passage from EPR into a metaphysical discussion of what physicists mean when they say 'reality.' This reformulation, together with Bohr's repetition of a few measurement procedures, has a strong rhetorical effect. Following Bohr's analysis of measurement procedures time and time again, the reader enters into Bohr's frame of mind and, without noticing, loses any critical perspective on the verificationist ground that Bohr gradually and carefully builds. By tinkering with the wording of EPR, Bohr creates an illusion that Einstein, Bohr, and the reader all share the same epistemological stand concerning the connection between theory and experiment. It is on this 'common' ground that Bohr 'defeats' Einstein. (Bell, p. 155)

The connection between theory and experiment (that Bohr subtly argues) is measurement as the sole determiner of existence. When chance relates to measured events, then Bohr is right and Einstein is wrong. What Einstein is trying to say is the chance of both events A and B are linked or entangled beyond the context of measurement. The chance of A adapts to the presence of B in the form of existence out of emptiness. A and B are both chances that have no firm form except in the temporal occurrence that realises existence through them.

Chance is a natural part of the way emptiness meets with existence. Chance or probability becomes for Bohr a concept about existence itself, as colour is a concept of optics for Newton. But this is to miss the very quality of chance. Chance is the orienting of events to show the whole potential of identity to the state of being lost. Chance happens all the time. When travelling, the getting lost in unfamiliar territory is continually opening up opportunity for chance to signify the way.

These chances are not mere happenstance, side-events to the main physical purpose of transporting oneself over the globe. The chances are the quality by which existence originally, uniquely and lightly shows itself to emptiness. Chances develop alongside other equally *bona fide* journeys to existence.

Chance is in scientific rhetoric, a slightly derogatory term that has to do with a quality that escapes proper definition. It lies forgotten at the bottom of the piles of papers talking about good definable concepts as matter and position. But actually what has been left hidden in the depths is the essence of life itself. The relation of existence to potential through chance is illustrated in *Figure 14* (overleaf).

Chance events are thus not exclusive in their occurrence. They are more the subtle communicators of existence. Chance events dance between emptiness and existence and so may sum the ground of potential into established fact in a coherent way.

The exile of chance into the world of concepts is further illuminated in the exchange between Sir Arthur Eddington and his student Chandrasekhar.

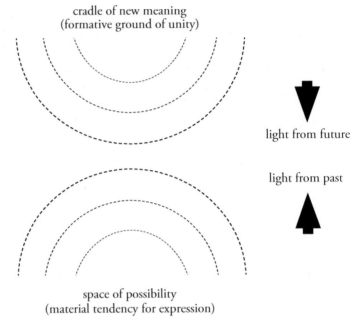

cradle of new meaning
(formative ground of unity)

light from future

light from past

space of possibility
(material tendency for expression)

Figure 14. Chance between freedom and resolution.

Naming and structure

The different historical understandings of *zero* in Eastern and Western cultures is illustrated by the difficulty of the great physicist and philosopher Sir Arthur Eddington to accept the theory of his gifted Indian student, Chandrasekhar. Eddington held to the belief of the ability of concepts to decisively frame the world. Eddington's faith in mathematics is illustrated in the following:

> Our whole theory has really been a discussion of the most general way in which permanent substance can be built up out of relations, and it is the mind which, by insisting on regarding only the things that are permanent, has actually imposed the laws on an indifferent world. (Eddington 1920, p. 197)

This understanding of physics by Eddington prompted Wheeler to say of one of his most talented students Peter Putnam, who fell from academia through some type of inner upheaval:

> Under the influence of Sir Arthur Eddington's *The Nature of the Physical World,* he had come to believe that all the laws of nature could be deduced by pure reasoning. Try as I might, I couldn't seem to disabuse him of this belief. (Wheeler 1998, p. 254)

This attempt to see the world through the names we give it, was the goal of science according to Eddington.

> The fundamental basis of all things must have structure and substance. We cannot describe substance; we can only give a name to it. Any attempt to do more than give a name leads at once to an attribution of structure. But structure can be described to some extent; and when resolved to ultimate terms it appears to resolve itself to a complex of relations. (Eddington 1923, p. 224)

One had to seek in the mathematics by reasoning alone for what behaved as the thing one wanted to describe. One could then identify the physical behaviour with the mathematical concept as equivalent, without having then to inquire deeper into the substantive quality beyond its mathematical representation.

The black hole that Eddington then fell down was the prediction by his student Chandrasekhar studying the implication of relativity, that dense stars would collapse and lose all their structure. These physical black holes seemed to violate the very principle of knowing the world by passively naming its elements. This led Eddington in 1935 to publicly ridicule the discovery:

> The star has to go on radiating and radiating and contracting and contracting until, I suppose, it gets down to a few kilometres radius, when gravity becomes strong enough to hold in the radiation, and the star can at last find peace. ... I think there should be a law of Nature to prevent a star from behaving in this absurd way! (Eddington 1935)

This put–down buried the theory of the Indian Chandrasekhar until many years later he was to win the Nobel Prize for the discovery! The physicist Arthur Miller concludes:

Chandra's discovery might well have transformed and accelerated developments in both physics and astrophysics in the 1930s. Instead, Eddington's heavy-handed intervention lent weighty support to the conservative community of astrophysicists, who steadfastly refused even to consider the idea that stars might collapse to nothing. As a result, Chandra's work was almost forgotten. (Miller, p. 150)

This interchange shows the contrast in emphasis between existence and emptiness of Western and Eastern approaches. The discovery of the black hole was thus a cataclysm at the very foundation of Western thought. How had the power of zero been usurped from reason and been turned into a cosmic emptiness?

Where thought understood the world literally, the emptiness of the black hole was simply an end. But Chandra, accepting emptiness as potential, could see even to the possibility that the death of the star would result in the new light of a white hole.

Unravelling

Critically it was chance that ended up showing the real meaning of quantum theory, as described in Chapter 1.

In 1941, the idea of the atom simply as matter had split into the extremely destructive potential of the atomic bomb device. Heisenberg and Bohr had become spokesmen for the two stories, of the Germans and the Allies, each holding one side of this incredibly destructive potential. The whole attempt to find certainty in the stable building block of matter had split apart into this situation in which they were representatives of these two sides involved in an ideological conflict.

The division of the world into concepts and imagination had turned into this huge potential for destruction, able to destroy the whole world and Bohr and Heisenberg were the two protagonists of this power that had been unleashed.

In using number to divorce matter from meaning, in trying to hide what science was really about, in trying to isolate matter, as if it could be seen as having nothing to do with meaning, circumstances unfolded to write meaning back into identity in this archetypal destructive way. Destruction was voicing itself through the very matter that concepts had isolated and separated out from meaning. In the very attempt to exclude meaning, to write it out of what science was about, the meaning still came through, but in this completely destructive way. It said, 'If you want to live without meaning, then meaninglessness will make its nature apparent.'

Test of the nature of chance

Ridiculously, the whole edifice of modern science is built on – chance. Barfield analyses the place of chance in Darwin's theory of evolution.

> What were the phenomena of nature at the time when the new doctrine began to take effect, particularly at the Darwinian moment in the middle of nineteenth century? They were *objects*. They were unparticipated to a degree which has never been surpassed before or since. (Barfield, p. 65)

> The phenomena *themselves* are idols, when they are imagined as enjoying that independence of human perception which can in fact only pertain to the unrepresented. If that is, for the most part, what our collective representations are today, it is even more certainly what they were in the second half of the nineteenth century. And it was to *these* collective representations that the evolutionists had to apply their thinking. (Barfield, p. 66–67)

> By a hypothesis, then, these earthly appearances must be saved; and saved they were by the hypothesis of – chance variation. Now the concept of chance is precisely what a hypothesis is derived to save us from. Chance, in fact, = no hypothesis. Yet so hypnotic, at this moment of history, was the influence of the idols and of the

special mode of thought which had begotten them, that only a few were troubled. (Barfield, p. 68–69)

Physics is also about chance. Probabilities are central to quantum theory. The whole edifice of the structure of modern science thus rests on no-one really looking at the nature of this word – chance.

Chance isn't something that relates just to the known world but it also relates to the unknown. What happens when chance visits us in some state of being lost, where things do not add up?

Chance is not by definition just an aspect of existence. As with the Copenhagen meeting, chance has a way of reflecting back to us the whole meaning behind our behaviour. It is necessary to disentangle chance from our mental construct of how the world works.

Embodying division

We all think that the quantum world is there. We all feel terribly important in it. According to Bohr's Copenhagen interpretation, it is solely the use of concepts as revealed through measurement that gives the quantum world its reality. Without us observing the world in its conceptual foundation, the atom would just fold in on itself, unable to recognise any more how it was supposed to work. For concepts alone infer existence through the act of measurement. Reassuringly as when we touch wood, we pronounce our daily concepts to keep the atom from falling in on itself.

We have built a whole empire on the view that it is our concepts that allow reality to work. Identity only holds by the constructs of our concepts.

We need to be clear on this. We do not use our cars just to get from A to B, but our cars (and mobile phones) give over our human significance to a concept. This is what one does. The car exists because our use of the concept 'car' brings it being. But this existence of the concept 'car' is not just a gesture on its own. The concept that can be measured is the philosophical basis of quantum theory. Our identification of the concept, as 'car' precedes the existence of matter that is consequence of the theory. In acknowledging the world built through atoms, as 'car', we are breathing life into existence as a whole, including us.

And of course such a superstition works after a fashion. We rush around in these cars adding to a mountain of conceptual exchange that contributes towards economic growth. We skirt over a surface that is necessarily human in its construction. We have so outgrown the superstition of belief that we are shocked to hear even a whisper that what upholds reality is a superstition of belief.

The Egyptian gods who held up the sky have been replaced by man-made, hard-core, scientific concepts. We can prove it. If we took away the concept, that idea inside our mind that we substitute for reality (and as a car drives us from here to there), then there would be nothing for measurement to say about the world, which would leave the world as a stuck record, forever waiting for the next...

What was the word?

Indeed the next word escapes us.

Beyond concepts

The world of concepts detaching from the reality that had seemingly been its origin, is not the end. When retreating totally from the world of concepts, a link to reality remains, there is a potentiality in the concepts that is still effective in giving future place in the world. As in Pauli's dreams, concepts from modern physics are not simply describing something actual (the configuration of particles), but are also describing a journey of the spirit.

In what follows in my personal story there is also an understanding about the use of concepts in science in general and especially modern physics.

In the telling of my journey using the same conceptual parameters as modern physics or in Pauli's dreams providing the framework for his spiritual journey, something essential is being touched upon. There is clearly some other avenue open to the way concepts are used in physics that takes one into, rather than away from, the light of one's human presence.

There is a dimension of darkness, of the unknown and of mystery, into which we can drop, where chance is the guide of a way, journey, process of individuation. This dimension of darkness is transparent to the relating of events of chance to illuminate a composite path of possibility for those who enter. The domain of possibility one enters in surrendering one's own certainty relies on a possibility and chance, not yet evident, to realise the illumination of a unity that lives in the happening.

There is something more basic in the relatedness of concepts that can guide the being to express finiteness with regard to wholeness. These concepts are not about *thought* prescribing for the world a predictable form; they are about the actuality of experience arriving at the far shore of existence.

The freedom of concepts is in their ability to mediate the knowing of meaning against the challenge of emptiness. So the true concept, elicited from the book of nature most skilfully, has within it both the key to destruction and the way to meaning. The concept mediates the world that destroys itself and thus obliterates the richness of experience, or the world stands as whole realisation of meaning through the concept.

The focus of happening has to reside between the *zero* and the *one*, able to reference both the *zero* of the east, and the *one* of existence of the Greeks. The only way to do this is to experience the *zero* of the east in the relation to *one* of the Christian God and the unity of matter inherited from the Greeks.

In-falling light

In the black hole example of Eddington and Chandrasekhar given above, it seemed as if the interrelationship of certain concepts – matter, space, time and light – had a chronic flaw in them. This flaw seemed at first glance to bring down the whole endeavour of physics as giving conceptual structure to the universe (and thus troubled Eddington into making the claim that the phenomenon should be outlawed). In the circumstance of dense packed matter, as in a collapsing star, all matter and context of space-time fall into a singularity of non-existence, the black hole.

Eddington was one of the foremost physicists addressing this quandary. In 1924 he found in the mathematics a convincing solution to the problem (alas, one whose consequences he could not himself believe in.) Eddington changed the reference elements of description from matter, space and time into a new mix of these parameters describing a light photon that was falling into the hole. The concepts no longer annihilated each other, but allowed for a holding description of the phenomenon. (See Misner *et al.*, p. 828–29.)

What is more, he could uncover another aspect of the phenomenon known as a white hole from the perspective of a photon of light, that was arising out of the hole, as mediator of re-creation! (See Misner *et al.*, p. 829–30.)

To understand the fall from existence, we can no longer deal with concepts as static elements, but have to relate them to the actual experience of something as light falling, or conversely arising from the singularity (or hole) of existence. What worried Eddington about this was that physics could no longer be viewed as an abstract exercise of giving structure to the universe. It required the experience of destruction and creation first hand, to render a meaningful account. Thus Eddington felt the need to snub Chandrasekhar's claim that such bodies could exist. But the evidence of the mathematics was that the riddle of the universe, through its fall and subsequent re-creation, could only be given description in the experience of something falling and re-arising through the hole of conceptual darkness.

Eddington's change of conceptual coordinates amounted to the ability of an exemplar of darkness, or a visionary of light, to hold a constancy by which the fall or renewal could know itself as a structured process of consequent steps. The exemplar or visionary held the fall, that the material deconstruction was not a terminal fate, but an ordered dynamic in preparation for the regeneration of the whole.

The discovery of the black/white hole dynamic exactly reflects the ground of this book. The fall from the fixedness that our conceptual map gives us, allows a consequent re-arising through light to re-illumine the whole sense of the universe in which we live. However we cannot hope to understand this experience, in a static way, but have to live its consequence in the path of the shadowing and re-arising of light.

The black hole cannot be a purely objective phenomenon, since it involves the dissolution of the conceptual basis of physics. It requires a journey that allows the perspective of light over the duration of the fall and the re-arising of light. In suspending my reality to report on this fall until arriving at new illumination, my being as a participant in reality was able to provide just this perspective.

Fundamental question

With my travels in Africa and the Alps, a sense of acquaintance with a fundamental aspect of the universe was born. I had surrendered myself to a question at the heart of experience. Yet the job to which I returned in Rotterdam, Holland, typified the sense that life was the mere mechanical logic that could be coded into a machine, my computer programming task. Here the office mentality was a concentration of the scientific view that the universe had developed by an essentially material accident, that later threw up life and the ability in humans to reflect on where we have come from. According to this hypothesis the only true guide to tell us how to live in the universe, was the fortuitous ability of our minds to conceptualise the material processes of our origins.

I then learnt that the relation to the underlying question also included negation, the no-saying to the universe, that the question might be illuminated newly.

This negation of the question, I experienced in a state of despair, where an elemental aspect of myself was unable to find voice. This 'no' to the question of the universe sounded in my very deepest existence, beneath any outside factors of conciliation. And what happened in this 'no' to the elemental question was that the very ground of the universe fell away. There was no outside or inside, background or foreground, world or distinction.

This state of negation of the universe was not as I first assumed some terminal dislocation. The question could be re-asked in a new way, allowing a different type of answer that included my soul's affirmation. The despair that made me travel through the overturning of conceptual description also was the way for a fundamental renewal in the address of the universe.

This was my own journey, to fall into the very isolation of the connection between *zero* and *one*, the vastness of potential explored in travelling and the insistence on names, labels, and compartments in the western application of Aristotle's *one*. An attempt was made to resolve the paradox of the East and West, India and Greece, in the Renaissance, through thought and the intellect. Now happening itself was to live the *zero* coming into new relation to the *one*.

Falling off the chair

One evening, in Dizzy's café in Rotterdam sitting with work colleagues, we were playing a game of lateral thinking, where one person posed a riddle and the others had to deduce the events preceding it that had resulted in this seemingly paradoxical consequence. My riddle was someone falling off his chair. After half an hour of clue searching I had no idea how I could admit that my puzzle was simply made-up, without an answer, when one of the players twigged, 'Are you pulling our leg?' 'Yes,' I said, 'That's why he fell off the chair!'

The paradox of this actual game, played in the depths of a crisis, illustrates something greater. Although concepts have fallen from their role of following the rule of the game of providing examples of lateral thinking for the others to guess, the spirit of my attempt was still alive. The unity of what lateral thinking actually signifies, lives behind the muddle I had put myself in and the joke having gone on too long for me to admit it was made up. Thus even though the conceptual hold of what I was doing in carrying out this deception had gone, this does not mean the game is given up.

What happens now is a real case of paradoxical lateral thinking. It dawns on one of the players that the answer might lie outside the very rules of the game itself. 'Are you pulling our leg?' At this point the unity of lateral thinking as an intuitive whole reality grasps the opportunity to answer, 'Yes, that's why he fell off the chair.' In other words, the unity of lateral thinking shows itself outside the very rules that limit what conceptual thinking and its practice through these examples is supposed to be about. The conceptual take on lateral thinking through examples is always going to be flawed. But when one accepts the limitation of lateral thinking by inventing something outside of its rules, then the unity can show itself as actual in the very defeat of everything that the conceptual was trying to establish. Thus in the laughter of understanding, the conceptual has on the one hand been duped by our complete upturning of the rules and yet the very attempt has lent itself to tell us what lateral thinking is in its unity beyond any conceptual example.

At such a moment, the darkness suddenly and momentarily inverts, as if reason itself has been turned on its head and has allowed light to enter. The understanding given in light is more than my own specific situation in playing the game. Something deeper is being communicated in the way that the conceptual game of reason itself is being turned upside down. The depth of seeing in emptiness gives to this specific reality of the game in the café, a universal lesson. Existence is able to judge itself anew in the foundation of what is valuable.

Orientation

While I travelled, lived, laughed and posed myself in the world, it was always as if my ownership of experience stopped at the *thought* of it.

Back in Europe I had entered into an office of computer programmers, which was itself a world of pure thought. Computer programming was a profession living in the ownership of the *thought* of the world.

Although my direct hold on existence had dissolved, there remained a potency in an inner freedom, that was able to seek in the world a new coherence out of the collapse of old orientation.

I was reminded about travelling by the excited return of a colleague Stewart, from four weeks travelling around Kenya. Stewart's account awoke in me the following:

> I climbed the mountain, then I took the bus there and there was that local market and then I went into the mountains again and saw that ancient settlement; now I want to go to that country to see their forests, to see that waterfall, to see the lifestyle of natives living under those conditions. You overtake yourself. You no longer sit there with it all going on above you. You don't get up just because the alarm's gone off, you don't catch the tram because that is what you do every morning, you don't just sit there working because the thing's got to be finished by Friday… You go THERE because it's good to feel that you've spent the last four days climbing to the top of the mountain, it's good to sit with someone and watch the sun set over the waterfall, it's good to see how the locals live without electricity, it's good to see the monuments built two thousand years ago in homage to their gods… You overtake yourself. You talk to people because you want to, they're good to be with, you want to hear what they've got to say. You don't talk to people just because they work in the same office as you. You don't sit next to people for months on end with all going on above you and remain strangers. (Franses 1982)

In this sudden reversal of emptiness to reflect a light from somewhere up above, of the coherence of a new freedom, something strange happened. A friend came into the café and engaged in a conversation about the reality of the spirit against the convention of the world. At one point she used the words: 'Never mind your debt to society, what about your responsibility to mankind?'

I had used these same words myself some weeks prior to my encounter with emptiness. Emptiness could reveal a path to existence that was shared. The surrender of the world of existence, as called by cultural norms, allowed that potential speaks itself across separate realities. The emptiness, beyond the personal, allows for many individual perspectives to constitute the universal nature of existence together.

Quality

Balancing the world with nothing, I picked up Robert Pirsig's *Zen and the Art of Motorcycle Maintenance* and asked Quality to judge the scale.

Originally, according to Robert Pirsig, the *one* in Greece had referred to skilful action, *arte*, the unity a person brought to the world through the quality of their engagement. This was the belief of the Sophists. At the time of Aristotle and Plato, when philosophy came to be written down, the *one* became idealised in the world of ideas of Plato and as a logical concept in Aristotle.

Arte turned into a conceptual value. The emphasis of the ground of existence shifted from the experiential to the conceptual. This tendency of abstraction was then exacerbated in the way Greek thought came to the West through Arab translation.

Pirsig similarly seeks a unity between his experiences in the East having been in Korea in the war and as teacher and explorer going to the roots of Greek thought in the U.S. His breakthrough is then to understand that quality as the bridge between East and the Greeks, *zero* and *one*, spirit and science, is to be found in experience.

Following his own madness, in being forced down the road of reason, his return to sanity is in seeing the notion of quality as extant in the world. Quality lives in the excitement of adventure.

It is the spark of happening. And from quality, the world of both reason and spirit follow. Quality is an event of seeing. Quality is the content of chance. Quality thus lives in the world of experience and does not borrow from the foundations of science or art but is the origin of both these undertakings.

Pirsig chances his own sanity to return to the essence of science and art as the forming of existence out of emptiness. Quality tells me that notwithstanding the fall of reason and the stranding of spirit, the world is always open to reveal its whole nature anew, to the one who journeys.

Egypt

Quality determined my course of action to take a month off work and hitch to Egypt.

Travelling in my thoughts down a line of known reference points, time becomes my capacity for action to create what might be ahead. History stops.

A TV crew asks me in a randomly chosen interview what I feel should be done to restore the pyramids and I respond, 'build something new'. Following the line of the Nile, peasants step eternally in their robes to break my inner monologue with the offer of tea. Staying at a hostel near Luxor, the outcasts of society reflect the day in the simplicity of the still life drawing of a boot.

Struggling with the bank of inner thought and throwing myself into the day of tombs, pharaohs, a skull found in the sand, I try to communicate this find and someone thinks I am talking about a murder and loses interest on discovering it to be an ancient loss. A group on departure turns the lights out, leaving me momentarily in the actual darkness of a tomb.

The tombs straddle life and death, their back west-facing to the Desert and Death, their front, forward facing, to the Nile and Life. Inside the tombs, drawings depict the life of the deceased, representing the glorious scenes of earthly light to the inner chamber sealed in darkness. The funerary jars hold the worth of the human's life to be weighed against the judging deity. Colourfully, the world illuminates aspects of the divine, straddling their light across the boundary, facing back the Dead upon the arena of Life.

The temple of Luxor at six in the morning, before the tourist coaches arrive. Sphinxes still stretch some of the two kilometre way to the temple of Karnac. One obelisk (and its absent partner in Trafalgar Square), stands at the gateway, where entering the Temple, one manoeuvres around a Mosque, built on the ruins. The Nile's blue colours, the crisp air of another cloudless day and suddenly everything is here. Time flows from that ancient place through me, the past slips quietly through the ancient temple stone, the secret that it too had been living for the same future, that nothing about the world was yet decided. Even as I stand, in my jeans and cotton jacket, the substantiation of its message lives in the reach of my experience. Held in the light of the sun, its cycle is still eternal, its conjuring of earth's mystery still faithful. Time no longer separates, but rather communicates a developing depth of task, as relevant to the tangent of today as of then. The historic dissolves in a common endeavour that bestows upon me an equality of task with this ancient core. (Franses 1982)

Barfield:

←———————————— chance/order ————————————→

Chapter 5

Potential faces expression: the arising

The fall from a static conceptual foundation of objectivity, uncovers a new dynamic relationship to wholeness. Goethe becomes our guide in translating into practical terms the consequence of this surrender. Wholeness is responsible for the dynamic articulation of the concepts, which in turn are the mediators of the unity of form, as expression of wholeness. Goethe's method actively researches the basic conceptual foundation, to allow the concepts thus found to reveal the form's meaning. Here the mediating role of concepts is applied to Goethe's study of plants, the bacteria colonisation process, the slime mould, the self-relation of the cell, quantum coherence of photosynthesis, the bee dance: they all illustrate the transformation of chance into a unified arrival.

One as expression

For the Ancient Greeks, (see the quotes of Aristotle and Domninus in the last chapter), the *one* was the unit before quantity, the *one* allowed something to be known as a 'number of...' It was only the uniqueness of the existence of one stone that gave any meaning at all to two stones, three stones, four stones... Number, for the Greeks, had no abstract existence, except as it applied to real things. In other words, concept and number were accessed through a specific experience of encountering the world as one.

So for instance a temple had measurements for its columns' height, breadth and depth, but this was only an example of the divine proportion that characterised the unity of such buildings.

Temples, having this relation to a particular unity out in the world, were quantified in size as a secondary exercise. Similarly Pythagoras understood the notion of musical harmonies, without any attempt at measuring sounds as wave frequencies. What he understood was the proportions between notes that sounded harmoniously. So music had the quality of evoking the unity that was already primal in the very notion of music before any piece was played. Or the arrangements of the planets in the solar system played the music of the spheres, each planetary orbit mathematically constructed about perfect geometrical figures. The concepts mediated the experience of the aesthetic unity of the world.

This was the Greek understanding of *one* as the inherent quality of things from which concept and number followed.

We can symbolise this as follows:

{concept} and {number}- - - - - >)

The symbol >) shows the limit where the process of finite rendition actively meets with the unity. The limit to the process of finiteness is the expression of *one* in the world. The {} brackets signify the definition of concept and number.

The Renaissance drew into the creativity of the mind, the very unity that had existed before in the world. The search of the natural philosophers was to seek the unity that existed behind number or concepts. Numbers became not only more real than the things they represented, but it was through number that the real unity of the world could be found.

For instance 'general analytics' attempted to find whether there was an order of reality behind that which we immediately see. Vieta, a key proponent of this school, called this a finding of findings. Can we find behind numbers themselves, an order that applies regardless of the entity we are addressing? In algebra, the substitution of general elements of 'x' or 'y' for things took away all ties with the *one* that was in the world. The mathematics of number said how things in general behaved.

Descartes took this type of thinking to its extreme. He suggested the only unity was that which could be reached through the mathematics of number. He completely denied any reality approached through the senses. He discounted the unity which was already extant. He put in its place the general quality behind mathematics of number as the true source of order.

Pre-unity

The realisation of Descartes' dream was quantum theory. Quantum theory finds a way of arriving at the significance of concepts through the mathematics. The concepts that serve to substantiate the act of measurement are given an entirely theoretical relevance. The physics of concepts actively creates the texture of the world according to how happening is observed or measured. The concepts are now entirely abstract elements to do with observation that allow the act of seeing principally to be effective. The concepts are taken as the fixed basis of the effectiveness of the interaction of observer with observed.

Thus Greek thought is entirely turned on its head. The concepts in Greece were given by the aesthetic relation of harmony, proportion, ratio, to the unity in the world. Now the unity of the world is entirely in the mathematics that predicates all that can ever be known about reality as it is revealed by measurement. The concepts are simply the neutral, mental representations that order the seeing of the world into a structure.

pre-unity
        (<– – – – {concept} and {number}
        (<– – – – {electron} {neutron} {proton} {atom} – – – – – – {world}

concepts give structure to the measuring of the world as an ordered whole

Thus the limit to finite existence is now provided behind {concept} and {number}! The ability of mathematics to shed the skin of the finite and symbolise the virtual realm of possibilities is called upon each time there is a measurement, to determine how the finite world of matter works.

Concepts and numbers are derived from a pre-wave of existence that gives the basis of any finite account of the world (through the principal act of measurement). And now the virtual world seems predicated on nothing but a binary system of 0 and 1!

All value has been removed, for the game of concepts and number that we play, while defining us, is totally lost to our impaired sensibility. Only a mathematician is able to appreciate the chances in existence to which all phenomena are ultimately reduced.

A complete about-turn had happened from the Greek externalisation to the western internalisation of unity.

Goethe

It is to Johannn Wolfgang von Goethe that we need to look, not a scientist by training at all, for how to use concepts dynamically.

The world's possibility is to be found exclusively neither in an aesthetic address of the unity of existence as the Greeks believed, nor in the intellectual abstraction of the generic law, as in science.

The polarity of possibilities between emptiness and fullness establishes the paradoxical foundation giving logical significance in how and what we see. Concepts in this approach are the mediating influence through the abstract potential of emptiness (pre-unity) into the expressed realisation of fullness (one as expression).

pre-unity   **one as expression**

(< – – – – – – {concept} - {concept} - {concept} – – – – – – – >)

(< – – – – – – {concept} - {concept} - {concept} = – – – – – – – >)

concepts as development

The concepts and quantifications of their values, instead of being the end point, are the finite means of development that translate emptiness into the unity of the one as expression.

The parts are 'passive/active' 'holding/transformative' qualities between the form as whole and the development as dynamic. The parts each have the quality of the whole in them. The parts are not distinct elements of a passive system. The parts actively lead through their individual statement into a greater whole. So Goethe was able to understand the parts as the transformative progress through which the whole form arises.

The part is neither founded on an innate unity as in Ancient Greece, nor is it a virtual attribute of the dynamic mathematics, as in quantum theory. The parts dynamically mediate the whole. The pursuit of the meaning encountered through the paradox of a part that is also a statement of the whole may be illustrated in the biology of the plant.

Metamorphosis of plants

In the classical view of science, the various static elements in the chemical construction of the cell are activated by conditions in spring to gradually transform the elements step by step from seed into the different stage of happening that by autumn has played out the whole life-cycle in regenerating the seed. The make-up of the plant can be distilled to its static elements, for the happening (time) is something treated as quite separate.

In Goethe's process, all the aspects of the parts, the joining dynamic and the expressed whole, are all disclosed in the process of engagement. One does not need to apply concepts as a theory, the concepts dynamically mediate the what? how? and who? of the whole process, illumined as it were from within.

Goethe identifies the organs of the plant as both the physical parts but also the dynamic elements of transformation. His book, the *Metamorphosis of Plants,* begins:

> Researchers have been generally aware for some time that there is a hidden relationship among various external parts of the plant which develop one after the other and, as it were, one out of the other (e.g. leaves, sepals, petals, stamens). The process by which one and the same organ appears in a variety of forms has been called *the metamorphosis of plants*. (Goethe, p. 115)

Form is innate to existence itself, form is not simply put there by the projection of a theory.

> Goethe spoke of the particular individual plant as being a 'conversation' between the living organism and its environment. This metaphor draws our attention to the plant's *active* contribution to the form which takes in specific conditions, emphasising the fact that the individual expression of the plant which we see is the outcome of the active response of the organism to the 'challenge' posed to it by the environment. (Bortoft 2012, p. 78)

The finite expressing of the plant in its frame is totally dynamic to its setting. As Goethe remarked when travelling through the Alps, his native Weimar plants had adapted, with subtle variations, into the counterparts found in the mountain meadows.

In being 'plant', freedom jumps over all finite considerations of what fits where, and so on, and understands each organ as a perfect finiteness of means. As the whole is experienced, so the organs present themselves as pure function. They are not functions of something, they simply are the finiteness which allows the experience of unity in the whole identity. The understanding Goethe draws from the plant totally agrees with modern genetic discoveries. Goethe views the stages of plant development (leaf, sepal, petal, stamen and carpel) as transformations of a basic internal freedom that is exercised in different circumstances to realise the plant differently. Modern genetics now corroborates this.

Theissen and Saedler writing in *Nature* confirm the functions that Goethe understood intuitively:

Goethe was right when he proposed that flowers are modified leaves. It seems that four genes involved in plant development must be expressed. According to this model, the identity of the different floral organs — sepals, petals, stamens and carpels — is determined by four combinations of floral homeotic proteins known as MADS-box proteins.

The protein quartets, which are transcription factors, may operate by binding to the promoter regions of target genes, which they activate or repress as appropriate for the development of the different floral organs. (Theissen & Saedler 2001)

The identity of the four organs, as Goethe intuits, is not externally programmed but the result of an internal freedom, responding to context.

The puzzling thing is, as one professor of genetics put it to me, how Goethe could have got it so right over two hundred years ago without the resources of modern genetics. The answer is that he did it by learning 'to think like a plant lives' through the practice of active seeing. (Bortoft 2012, p. 62)

The leaf sequence orders the various organs of development as they appear in the life cycle of the plant. The framing of the whole nature of the plant is recreated by following the temporal sequence of the leaves in the cycle of their appearing as in *Figure 15* overleaf.

When one follows the cycle forward in time, then it seems as if the whole nature of the plant is allowed for by the development of the partial stages. One gets an impression of a development working to restore unity through the subsequent choices from the first breaking of symmetry in the original leaf. This is the *Light in Time* movement, where the happenings of time determine the stage of realisation.

Figure 15. Leaf-form sequence of sow thistle,
(Bockemühl, p. 5).

On the other hand, one can follow the cycle differently and go anticlockwise from the produced seed to the first leaf and receive the impression that the coherent unity, at the completion of the process, was in fact the originator of the dividing of wholeness that happened at the beginning. In this way of seeing, everything that happens in the living journey of the plant exhibits the character of the finished journey. If one practises this with the leaf sequence, one indeed experiences the whole, telling its story back on itself. One feels then the black eye of the poisonous plant, or the smile of a daisy, as if these characteristics were the object of the growth, now found in the world. One meets with the character of the plant that becomes a subject retelling its existence through a certain cycle of nature. This is the *Time in Light* movement.

The freedom of expression in the parts is open for the imagination to envision the whole that coherently signifies each of the parts. The form of the various leaves, up to the flower organs, has an ambiguity or freedom, that allows for a whole to signify all the parts relatedly. This is the nub of what Goethe means by pre-unity and one as expression. These unities are not there before the plant journeys into being. The whole signifies the parts dynamically to relate them together as aspects of a single meaning.

In following the round of the leaf sequence, the whole appears as inevitable. One finds the character of wholeness, either by going forward into the actual physical story by which the whole arose out of the parts; or backwards into the way a unity of self-generating potential preceded the parts. The parts as a way to unity may tell the whole in two ways:

1. as an actual physical occurrence through the developmental stages of leaf, sepal, petal, and so on;

2. as enfolding into the parts, a nascent unity from which everything develops. The pre-unity of potential is not abstracted before the parts, as a mathematical theory may claim, but is imaginistically envisioned as the unity that holds the parts related in a single meaning.

What Goethe is doing is not trying to abstract the plant from its context, but embed the plant into a life process, as the dynamic source of being. Goethe seeks the quality behind the various forms of the plant.

> Here we would obviously need a general term to describe this organ [ur-organ] which metamorphosed such a variety of forms, a term descriptive of the standard against which to compare the various manifestations of its form. For the present, however, we must be satisfied with learning to relate these manifestations, both forward and backward. Thus we can say that a stamen is a contracted petal, or, with equal justification, that a petal is a stamen in a state of expansion; that a sepal is a contracted stem leaf with a certain degree of refinement. (Goethe, p. 120)

Goethe then went further into seeing that in studying the difference between all plants, one could allow all these partial forms to suggest the unity that related all these variations into the coherence of the family of 'plant'. Note that Goethe does not pursue this goal intellectually. Instead he allows the imagination to see the necessary unity of which all plants are an expression.

> A challenge hovered in my mind at the time in the sensuous form of a supersensuous plant archetype [*Urpflanze*]. I traced the variations of all forms as I came upon them. What passionate activity is stirred within our minds, what enthusiasm we feel, when we glimpse in advance and in its totality something which is later to emerge in greater and greater detail in the manner suggested by earlier development. (Goethe, in Mueller, p. 162)

The abstract unity is not arrived at by Goethe as something which reason has conquered as a map of the territory of existence; the abstract unity is touched by the fertile imagination as the backward discovery of origination. The movement to the source knows the journey backward through the manifest forms, to the spring of origination in which the transformative passage to expression is couched.

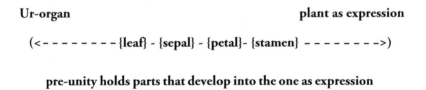

Ur-organ   **plant as expression**

(<- - - - - - - - {leaf} - {sepal} - {petal}- {stamen} - - - - - - - ->)

   pre-unity holds parts that develop into the one as expression

Urpflanze   **kingdom of plants**

(<- - - - - - - - {plant} - {plant} - {plant}- {plant} - - - - - - - ->)

Method

The Goethean methodology unties the rational, the imaginative, the intuitive and the aesthetic aspects of seeing, into a collective encounter with the whole nature from which all these partial aspects of perception arise. The seeing is not limited to any one partial aspect of human perceiving, say the scientific valued over the artistic or the aesthetic over the factual. Nor does the methodology of Goethe prescribe an instruction for how to blend these different ways of seeing together. Instead the whole process gives coherent meaning to the gleanings from all these different ways of seeing. The method sees into the meaning that is the origin of everything that is observed.

A freedom in each aspect does not close the seeing about one particular interpretation of what is there. Instead the openness of the inquiry allows the quality of meaning to signify the relation between all the aspects of seeing in which one was engaged.

Goethean seeing begins very much like modern science, by looking at the plant or the tree and drawing it and noticing and recording, very neutrally, what is there. For instance, looking at a chestnut tree in spring, one observes the way last year's buds unfold this year's growth. The buds shoot forth into new leafing branches. Some buds, instead of extending outwards, develop into flowers. What one at first thinks of as quite mechanical, becomes in looking, quite magical. For instance it is very hard, in looking at the closed winter buds, to imagine the growth of spring that is to occur through them.

The next stage is to fill one's imagination with this process of development from the bud through leaf and flower. Instead of imagining what is there in terms of one's own concepts, one sees into the quality of gesture that the chestnut realises.

So the first question one asks is, *'What? What is there?'* The second question one asks is, *'How? How is that tree being?'* One finds a sequence of steps in which the plant or the tree develops. Even though one has represented a snapshot by drawing exactly what's there, the snapshot isn't the thing itself. What one wants is to find that film, that moving series of snapshots, which is the 'How?' of 'How does that whole embody itself through the parts?' One tries to get various perspectives. The first question 'What?' gave a single perspective or snapshot, but one tries to build up a series of snapshots, to make a film. One might say, 'Here's a huge chestnut tree. But now I'm going to look at a conker seed. And then I'm going to look at a little tree, like this, and then a slightly bigger tree.' One tries to form many snapshots, because one wants to experience this dynamic quality of the tree.

The last stage of the process, called Seeing in Beholding, is asking the question 'Who are you?' Here's an example: one might be busy trying to understand something, and one thinks, 'Well how did this fit with this, and how did that fit with that?' And then suddenly one sees the whole journey. Then one really knows the whole, because one can see how it developed. One lets one's imagination go with all these stages of the journey. And one tries to let something fill in the gaps. So one has all these stages that one has looked at or seen, all these different perspectives and one lets the film run through one's imagination. Suddenly then the elements observed in stage 1, the happening in the film of snapshots in stage 2 and the seeing of stage 3 are part of a single unique unfolding that is showing itself. And that tells how the whole quality of the plant lives in those parts, those snapshots.

One is travelling in one's imagination the opposite way to the plant's motion of expression. One is returning to the pre-material unity that precedes the distinguishing of concepts. One allows this journey of parts to the pre-material unity to distinguish itself in one's imagination, by a movement experienced as putting together the various conceived parts.

One allows the snapshots of existence to form in one's imagination and the internal dynamic to identify itself. As with the leaf sequence, one feels the whole that lives by and through the parts.

The concepts are not understood statically, but dynamically as they point to a meaning through them.

Re-membering

Goethe understood that in wholeness, all of time was contained within itself, so that the process of looking at a plant contained within it the secret of how the plant became itself. One had to offer to time a question so that the meaning revealed to the seeking was one with the meaning in the world of the plant. In the process, the nature of the story that is told becomes one with a meaning that needs expression in the world. By opening up to a connection of science with spirit, physics readmits an active living quality, seeking elucidation. This is as with memory, that we do not store old events somewhere in our brain, but we allow that an openness overlays a current questioning with the past experience. The basis of Goethe's method is that we have to actively establish this bridge of seeking through darkness, so that the light that is the plant's active meaning can visit itself upon our questioning.

I am drawn back to Eddington's coordinate system that he used in 1924 to penetrate the black hole phenomenon. One could look into the interior of the black hole through the fate of an in-falling or out-going photon of light. But this made no sense to Eddington, as it appeared to dissolve the ability of the scientist to abstractly conceptualise what was happening. In Goethe's method, that in-falling or out-going perspective of light is the patience to surrender one's participation until receiving the illumination of meaning. The phenomenon is written into the attitude of time, as the dynamic receptacle of light's showing of meaning.

What troubled Eddington was the puzzle that a certain type of dynamic observation was required to give conceptual meaning to the objectivity of what was there. But Goethe (of whom Einstein was a great advocate) understood practically that to apprehend the process of origination, one had to make oneself vulnerable in time. As light gives structure to relativity, so light is part of the reality of the way the world described by relativity communicates its meaning. Only when our perception is up to speed of light, as participant in the revelation of meaning, does the dynamic instance of relativity tell us its deepest secret.

This changes the very nature of our perception. Instead of seeing through a map that our mind has synthetically constructed, our mind is more an instrument that tunes in to the fundamental need of existence. Perception is more of an activity, a 'remembering', that recalls the originating process of the organism, as a living revelation of existence.

This nuance of relating was also brought out by Goethe's very public disagreement with the teachings of Newton on light. For Newton, light was just another material thing, and the colours were the mechanical parts on which such a description was founded. But Goethe interpreted the world as a dynamic process, in which colour was the boundary of interaction between dark and light. For Goethe, one had to know the living quality of light, as a received revelation, to then give place to colour as a concept of existence. Goethe's world was alive with the creativity of the artist and the communication of meaning. There was no separation in Goethe between the practice of the artist to depict a meaning, and giving to light a place as generator of scientific order.

The paradoxes in different ways of conceiving the world come together in the meaning of 'seeing'. The concepts do not appear to us statically as either this or that. The remembering, the seeing, the concepts, the unity, are all one. Remembering and foresight are both there in the phenomena. We have the insight into the abstract origin of everything that has been as a remembering, at the same time as we have the foresight of the whole unity, of what the development can become. Everything is woven through each other until the unity places everything in fulfilling relation.

The confusion in the western scientific mind is that having understood the way through concepts back to the abstract unity, it imagines the concepts are purely a mental construct (or else the structure of what is there). In Goethe's approach the concepts are not static elements the mind holds as the intellectual key to reality. The very foundation of our understanding of the world abandons the finite, as a statement in itself. Science is now a speaking of a journey, where the arrival at whole meaning, transcends but also determines the capacity of space/time description.

Genetic meaning

A huge amount of evidence on genetic involvement has been drawn from biological investigation. It was originally assumed that genetics would provide a passive fingerprint of how randomly coded molecular sequences had encapsulated certain fundamental instructions for the build-up of the cell. The goal of biology was simply to unravel which sequences applied to which behaviours, allowing us to correct imperfections that could be discovered in certain individuals. Indeed certain diseases have been identified as resulting from mutations in the copying of the gene through inheritance.

However the overwhelming impression is that the genes are involved in the most intricate of dances, where hundreds of genes are able to produce proteins that activate or repress other genes in a dynamic of huge complexity. The idea that this dance is genetically pre-programmed diverges from the evidence.

Even silencing various key genes can often result in a totally different genetic pathway being created to realise a vital behavioural action. The genes, the proteins and the cells are seen to be dancing a tension between potential and expression, responding in their mediation to channel the potential of something actively at play into the form of that which is expressed in its own autonomous identity. The dance of genes, proteins and cells, exactly mediates a process of unity resolving itself into form. The complexity of interactions at every level is the process of figuring out how unnamed possibility constructs a frame for its own expression.

Collective expression

Ben Jacob illustrates the individual actions that live a simultaneously coherent unity in the case of bacteria colonisation.

Bacteria were the first life form on earth and though able to exist independently, they form colonies of 10^9–10^{12} bacteria. What again is at the very heart of these activities, is the synchronisation of temporal activity so billions of bacteria can act with uniform behaviour. It is the trigger of environmental threat that sets off bacteria into this collective form. (Consider organising twenty humans around a table to agree on something!)

Ben Jacob summarises this universal organisation:

> We are referring to a sense-based generation of meaning that occurs at all levels of an organism's hierarchy of function. Meaning requires on-going information processing, self-organization and contextual alteration by each constituent of the biotic system at all levels. The macro-scale selects between the possible lower scale organizations that are in an entangled state of different options. (Ben Jacob *et al.*, 2006, p. 518)

The last phrase 'entangled state of different options' might be translated in non-physics language as 'woven into their collective potential for meaning.'

The colony identifies itself through a process of discovering the single meaning that establishes the whole as the element of existence beyond the individual membership. It is this organisation alone that lets the individual bacteria survive. The future communicates the single option of coherent assembly that enables survival.

Ben Jacob then draws the parallel with this type of meaning-making with quantum theory in physics.

> Metaphorically, the above picture is similar to the notion of quantum mechanical collapse of a superposition caused by measurement. There are two fundamental differences however in the selection from an entangled state of options: 1)... In the organism's decision-making (selection of an option) an external stimuli or received information initiates the selection of an internal specific option. 2) Both the external information about the stimuli and the selection process itself are stored and can generate an effect on an array of new options and consequent selection processes. Therefore the unselected past options are expected to affect subsequent decision-making. (Ben Jacob *et al.*, 2006, p. 518)

Space and time are to be seen as the mediators of relationship, in which the internal commitments to change are configured to allow the meaning of 'colony' to be expressed through the individual actors. There is no need to interrupt this process in order to measure with rods and clocks exactly what space and time signify in their internal dimension. The origin of the concepts of space and time are the engagement of the individuals with their collective meaning. The essence of *colony,* which is encountered, lives in the arraying of individuals in space and time that allow the whole expression.

The bacteria effortlessly switch from a satisfaction with their own independence to a seamless fitting of their functional capabilities to serve a colonial organisation. The finiteness of their meagre capacity as single-celled organisms instead of being opposed to the cause of their unity, allows them to simply live their means in the unity they now adopt. The simplicity of means is not contrary to the attempt at unity but allows it to become.

The effect of a physical influence, say an electromagnetic field, as Ben Jacob describes, is to alter the response to a threat stimulus by changing the colonial form. In other words, the finite circumstances play through the adopting of unity to shape with context the particular expression of the whole.

This interrelationship of individual commitments and whole action is further evident in the act of sporulation. When the bacteria are unable to find food, they are able to commit their individual existences to alter the whole state into one of sporulation, or drying out their cells, to wait in spore state until more fortuitous circumstances arise. Ben Jacob discusses thus:

Sporulation is a process executed collectively and beginning only after 'consultation' and assessment of the colonial stress as a whole by the individual bacteria. Simply put, starved cells emit chemical messages to convey their stress. Each of the other bacteria uses the information for contextual interpretation of the state of the colony relative to its own situation. Accordingly, each of the cells decides to send out a message for or against sporulation. Once all of the colony members have sent out their decisions and read all other messages, sporulation occurs if the 'majority vote' is in favour. (Ben Jacob *et al.*, 2004, p. 368)

The wonder then is not that this internal barter between billions of bacteria and their collective decision should occur, but how the inner potential and outer actualisation in the simplest of single celled organisms, informs the space and time of their experience with a collective action.

It is ridiculous to imagine such behaviour as determined by an external organisation as a system of number in which each individual bacteria is individually directed. Instead it is enough that the association of bacteria mark each others' presence as significant in contributing to a realisation of their whole completion. The bacteria get the behaviour of their fellows, which before had been out of vision of their own individual lives, in order to found a future unity of completion, upon the basis of this collective holding of potential.

pre-unity   **one as expression**

(< – – – –{bacteria} - {colony} - {sporulation} - {bacteria} – – – –>)

Development of life

Crucial for the development of life was the way the original bacterial organisms inhabiting the earth were able to use complex chemistry to obtain energy from their surroundings. The task they faced, without any chemistry degree or lab in which to practise, was how to profit from imbalances in the environment and early atmosphere in order to extract energy to fuel life. In principle, this process is almost impossible. According to the second law of thermodynamics, energy tends to go from ordered states to disorder. But life is conjuring up from disorder, the order of the organism as seen in complex metabolism. Moreover this conjuring trick of exploiting the environment for energy was solved not once but many times by living processes themselves, as the atmosphere changed. Finally photosynthetic bacteria cracked the hardest puzzle of all, obtaining energy from carbon dioxide, water and sunlight. To do this, they combined two earlier transformation processes to create a cascade of opportunistic reactions, leading eventually to the capture of energy into the inner electron flow of the organism.

The feat of arriving at such a complex manipulation of chemistry without any lab in which to experiment, demonstrated the fluid dynamic way in which the basic concepts of manipulation, the cell, the proteins, electromagnetism, are not inventions of man's mind, but are the dance nature herself utilises. This understanding is further brought home in the discovery in 2007 (Engel *et al.)* that photosynthesis does not work in a mechanical way, transferring one photon of captured energy at a time, but in a entangled way, exhibiting a quantum beat of coherence. The coherence amounts to each act of transfer sampling all possible routes of excitation, before finding the path that best fits with the efficiency of the process over the whole. The result is an astonishing 95% efficiency rate of energy transfer, a startling achievement, far greater than comparable human devices can achieve (currently around 43%)! A wave-like coherence in activity is apparent over the system as a whole. The process exhibits a beat through time, which demonstrates a collective synchronisation of individual actions to harness energy.

The potential for collective meaning in the transfer of energy down one particular molecular pathway involves the simultaneous sampling of all states to determine a single individual course of action. In other words, whole and part, as is always the case in quantum systems, are inseparably interlinked. The individual act of the transfer of sunlight makes its contribution within a transformation of the collective state. The usual quantum paradox applies here. The whole and the particular both dynamically realise the other.

However there is a key difference here. The quantum correlations do not apply to an entity already existing. The efficiency of the transfer of energy gives definition to the photosynthetic organism. The coherence of quantum processes is rooted in the emptiness of potential out of which existence comes. Emptiness is able to join the disparate potentials on the edge of what makes life into a defining activity for the realisation of the whole. The photosynthetic process is engaging with the life/death process, where, at the edge of annihilation, molecules pool their collective touch with destruction into a vivid realisation of life.

It should not be underestimated what a feat it represents to efficiently tap (from sunlight, carbon dioxide and water) enough energy to fuel the organic process. Even a technological production developed by the most sophisticated of man's inventiveness cannot achieve an efficiency that can even compete with the efficiency of the natural process. As seen above, the organism achieves a 95% efficiency at the very edge of what defines it as existence.

pre-unity   **one as expression**

(< potential organism --- {synchronous excitation state} --- energised organism >)

Not only does life solve the dilemma of making order from the disorder of the random state of the environment but it does so with an innate elegance and efficiency that is almost perfect (that is, 100% conversion of energy). The organism does not stumble accidentally across a boundary of form. It launches itself out of disorder into an almost perfect statement of whole organisation.

The cell relates to itself to become itself

The biological cell spans the entire evolution of life. The cell marks with a membrane, a territory of distinction inside. On the other hand, the cell only has meaning when seen for its ability to divide. That is, in the act of division, the cell is able to know itself (the divider cell) and to see itself as another, a copy of itself, as separate to itself (the divided cell).

There is in a sense only the first cell, that has divided itself many times, or the cell is the self- relationship whose meaning is the many cells.

The cell has a meaning because at the origin of life it was able to represent existence and found a way to make that existence again through its own living. Somehow it took what it was, this living thing, and found a way of making that living thing again. If the cell had never replicated, it wouldn't be a cell. The basic relation is then shown in *Figure 16*.

Figure 16. A cell relates to itself (as a copy of itself) to become itself.

What we understand by a cell, is not just one thing but an endless proliferation into many. Part of what a cell is, is its ability to reproduce what it is.

The problem of going into the cell as a thing in the lab and taking it apart into genes and proteins is that one is looking at what the cell has become, out of the context of relationship. Modern systems biology is moving to the idea of relationship, but is holding on to the idea of the cell as an entity, as an elementary building block.

What makes the cell the cell, is the ability to relate to itself. In the act of splitting itself, the cell takes its identity and brings this identity out and completes itself in how it has made of itself a copy of itself. So the cell is the cell through life knowing itself and forming a relationship through what it is (into another one of what it is). From the beginning the cell was as much the ability to reproduce as any sole properties of the thing itself. The cell knows itself to become itself again. Everything about a cell, every property or characteristic of the cell, derives from the fact that first of all it is able to make itself into a copy of itself. Everything a cell is, starts from the fact that it is able to relate and in that relating it becomes everything it is.

The cell or any other element – an atom, a gene, a protein, a self – lives by knowing itself in relation to itself. The cell's existence is not derived from its parts. Everything about the cell, its relation to the world, the relation of the whole to its parts, its relation to its self – all are included in the entity that knows its journey through the world as whole. The cell is an identity that takes up into the quality of change, a reference out of which life can orient its subsequent development.

The bounded existence doesn't produce as a factory, something less than itself. It is able to reproduce exactly what it is, the whole entity. The cell stands in relation to something that is as real as itself. Its reality is totally defined by its relation. Everything that constitutes the cell is reproduced, so you can't say what it is except in its ability to reproduce itself. The relationship in which the cell becomes the cell is the very nature of what defines it as a 'cell'. The relationship is *the* essential part of what it is.

If you look at the cell and some of the properties that have evolved, for instance acquiring a nucleus with DNA as the basis of multi-cellular life, then these are new ways of the cell becoming the meaning of the relationship to itself. All the qualities it has, all the ways it sustains life, all follow from first of all reproducing a new copy of itself. Life expresses the same but differently as another. The relationship is able to understand itself in becoming another cell and another cell...

The relationship is paradoxical and changes the way we understand existence. There is no such thing as a cell out of its context. The cell is a cell by courtesy of the way it relates to itself to become itself, the cell in having this relationship to what it itself is, is always being born anew.

Identity and relationship arise together

Cells relate to other types of cells in order to find through these relationships, their own meaning of identity. In the same way the proteins and the genes that make them in the cell are involved in chains of relationships (proteins are able to activate and de-activate the production of other proteins) so that any one function of the body actually has a whole chain of active proteins involved. There is something that needs to relate to become itself and it does so by relating through other things, to make a network of relationships that end up identifying this thing through a whole lot of intermediaries in between, by which it becomes itself. Underlying a network of relationships, the identity is not a thing that has a relationship, but an existence that only is because of its relationships. It has to be in relationship to be itself. A cell has to be in this relation, to that relation, to this relation to be a cell.

pre-unity   **one as expression**

(<– – – – –{cell} - {protein} - {gene}- {gene} - {protein}- {cell} – – – – –>)

Life isn't stuck to one particular form. Who we are is in our ability to make ourselves anew. We can know ourselves as life and we can understand ourselves in that knowing from our experience into something new and we recognise ourselves again as who we are. It's totally new. We've never done it before, we've jumped out, we were mathematicians and now we're travellers. We do something totally new and it completes our relationship of who we are. We recognise who we are even though there is nothing similar in that new place. It completes our relationship to who we are. In fact we are more ourselves in having put ourselves out and made ourselves different.

So each time the cell makes a new type of cell, it is more than it was. Each time the cell becomes itself differently, it becomes more secure in what it is. We don't need permanence, for our very existence is finding ourselves in relation and through that becoming who we are. The only way we can go is into the unknown, finding ourselves again and that's all that the cell does. The cell doesn't have a fixed existence. It finds itself in relation in different ways, in different contexts.

Mediation of forms

The slime mould gives a slightly different nuanced view of the colonisation process. The slime mould consists of individual amoebae that have autonomous lives of their own, until food scarcity draws them together. The call to unity now results in different roles being taken by the amoebae. A group of amoebae initiate the call to join together in a unified state. The leaders then form the head of a composite body called a slime mould which has all the characteristics of a single organism. It moves rather like a slug, with no sign that it is in fact composed of cells that are autonomous individuals. When finding an appropriate place, it forms a fruiting body. The initial leaders of the movement sacrifice themselves to form a hardened stem, while the other cells sporulate off branches of this central stem. This tree-like structure persists until conditions are again providential for life.

Once more we see here how the commitment to unity instructs the finite behaviour of the individual cells. The individual cells make choices, which are not confined to their own benefit. The behaviours thus lead outward from the individual into the expression of the unity of the pseudo organism, the slime mould. However the slime mould shows that there is no distinction between the individual actors and the whole organism they form. The individual amoebae could as well be the fixed cells of the organism. The behaviours by which the individuals know themselves in relation to the whole, are at the same time the unity organising itself through the individuals. In this latter regard the slime mould is no different to an organism made of cells.

The unity imagines itself into the necessary behaviours that allow the whole form to take on a temporary existence in the world for the sake of regeneration.

pre-unity   **one as expression**

(< – – – – {amoeba} - {slug} - {stem} - {fruiting body} - {amoeba} – – – >)

The bee dance

The individual bee does a waggle dance which communicates to the hive collectively, relative information such as the direction and distance to patches of flowers yielding nectar and pollen. The bee does a performance, which abstractly distils the context of the bees' existence to a communal strategy of further exploration. The bee is a riddle standing between self and its action.

{abstract} self = self {expression}

The insight of the bees agrees with Goethe's approach to the plant. The conceptual relation of the bees' relation to the location of the nectar is not a mental activity. The concepts of relation to the location of the nectar may be acted out in an abstract way into the code of a dance virtually conceived. This information gathered from the bees may then be lived forward into an activation of the navigation skills of other bees to the flower sites.

The concepts are not intellectual. The compass of bee activity that allows the bee to actively navigate a direction may be followed back into an abstract distillation of the essential map of the nectar's presence. Concepts, as finite mediators of wholeness, have two expressions: they may be activated as physical skills of navigating the countryside in search of nectar, but also they may be questioned for their abstract significance.

Concepts are about the divination from the evidence of a direction that points in the way of the good of the whole.

The hive does not consist of a collection of bees who then sit down and work out a code of identification and communication of nectar sources, even many kilometres away from the hive. The activity of the bee dance is the action that defines the collective existence of the hive.

pre-unity   **one as expression**

(< - - - - - {nectar} - {flowers} - {dance} - {nectar} - - - - - >)

Caring about the world

The mistaking of conceptual reality for an intellectual understanding was brought out by a dialogue with a colleague. His stance was that reality was understood by reason alone. Angered by a loose comment implying the spiritual path, he set out to demonstrate by reason his viewpoint of rationalism.

My colleague wanted to establish once and for all the empty rational endeavour of life. As he honed in to making his point, the horizon of possibility seemed to visibly shrink, until what seemed a moment of total dissolution approached. Suddenly unity filled my heart with quite a different meaning from the situation. His wave of aggression broke against something solid.

> Perhaps the feelings on a country bench
>   That the world was real, and I was too,
> Or the sight of a moon, full and bright,
>   Its eyes turned away from the beauty-scarred earth

In giving myself up to the emptiness of approaching annihilation, the conceptual understanding of my experience in the world, re-formed about a whole communication of what it meant to live. The conceptual instead of breaking the world apart into divisions, gathered itself up at a centre of a map of making decision.

'You can either care about the world or you can't.'

The paradoxical frame of existence can only be tested in its living. Everything that is to happen through the acceptance of paradox is to be understood from this basic dichotomy that lives at the foundation of experience. At the same time, the insight is more than just an intellectual understanding. The paradoxical dichotomy also establishes a reference for the gathering of experience into the order of a defining meaning. The insight thus is something dynamic in my heart, already filled with the potential of event, to arrive at a unity of meaning.

The context of my journey is able to concentrate in me sufficient knowledge of my direction, to broadcast a sign, to alert others to the universal endeavour. An incompleteness of being draws out a symbolic form in which further development of a journey can happen across individuals. The signifying sign is transparent to the active meaning that later can be made through it.

pre-unity   **one as expression**

(<- - - - - {world} - {care} - {choice} - {uncare} - {world} - - - - ->)

The world reveals itself through the whole unity of what can become. The sign is a whole pointer from a partial state of realisation. The lines of realisation and negation (for both are allowed for in a journey that is yet to define itself fully) develop the whole sign so that the destination can bootstrap itself up from nothing. The realisation and negation lines come together in giving to the partial information, enough content to transmit and receive a common knowledge of passage.

'You know not everyone thinks as we do.'

Goethe:

potential/expression

Chapter 6

Emptiness faces form:
the hub and the rim

In the last two chapters we have traced how that concepts are not static elements of a thought landscape but dynamic aspects in the synchronisation of form. Using this insight, the mathematicians Spencer Brown and Lou Kauffman transformed the static elements of numbers into dynamic correlates. The basic movement of the world is in the logical elements themselves.

Paradox enables the appearing of meaning from within the process of engagement itself (rather than imposed as an understanding from outside). The paradoxical work of Spencer Brown and Kauffman surrounds the failure of mathematics to stand in its own finite statement.

Redeeming arithmetic

Potential is everywhere. Existence gives finite boundary to something. This is depicted in *Figure 17*.

Figure 17. Potential in relation to existence.

The novelty of Spencer Brown is to identify the act of crossing from potential to existence (or existence to potential), as the primary act of this world. This primary act of crossing is given by ⌐. This is shown in *Figure 18*.

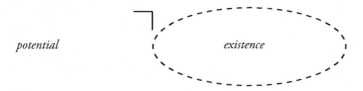

Figure 18. The crossing between potential and existence (or vice versa), with Spencer Brown's symbol.

Existence is only entered through the sign ⌐. There is existence in the crossing from what is without form into something. Thus the mathematics is based now on an action ⌐. Imagine the symbol as a vertical line of identity joined to a horizontal line of creating in time, a place in emptiness.

The separation of *zero* and *one*, is resolved in the dynamic relation of a potential (empty space) to the *one* as existence (the act of crossing). *Zero* and *one* are not abstract identities connected by arithmetic rules. *Zero* and *one* are unbounded and bounded space that relate through the act of distinction as shown in *Figure 19*.

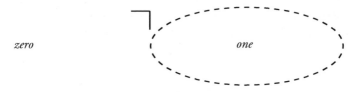

Figure 19. Relating zero and one through the act of crossing.

The basis of arithmetic becomes the following relationships between unbounded and bounded space. (The reader need only feel the movement in the mathematics or can refer to the simple explanations of moving between unbounded and bounded space.)

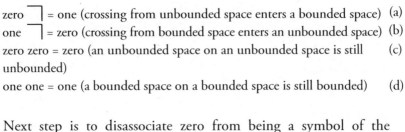

zero ⌐ = one (crossing from unbounded space enters a bounded space) (a)

one ⌐ = zero (crossing from bounded space enters an unbounded space) (b)

zero zero = zero (an unbounded space on an unbounded space is still   (c)
unbounded)

one one = one (a bounded space on a bounded space is still bounded)   (d)

Next step is to disassociate zero from being a symbol of the mathematics and just to make it nothing at all, an empty space. Zero is removed, making the following:

⌐ = one (crossing from unbounded space enters a bounded space)   (a)

one ⌐ = (crossing from bounded space enters an unbounded space)   (b)

Having made zero an empty space, we now have the identity that:

⌐ = one   (a)

Making this substitution we then have the two axioms of the new arithmetic, overturning everything we learnt at school about 0,1,2,3 … addition, multiplication …

⌐⌐ =   (b)

and ⌐ ⌐ = ⌐   (d)

With these two relationships, we are ready to start on our journey into a redeeming arithmetic. The *zero* from the East and the *one* of the Greeks had been separated for hundreds of years in western mathematics. The relationship of *zero* and *one* given so simply above (without worrying excessively about the mathematics) is the starting chord for a new music.

The tension of the *zero* of the East and the *one* of the Greeks is now finally addressed, in the meaning that comes to resolve the tension of opposites.

What is accomplished, in mathematics, is a transcendence from a given state of vision to a new and hitherto, unapparent vision beyond it. When the present existence has ceased to make sense, it can still come to sense again through the realisation of its form. (Spencer Brown 2009, p. *xx*)

Quality

What has happened is quite shattering for the foundation of a world that prides itself on adhering to logical principles. We are accustomed to see the system of numbers as isolated elements that relate through the rules we learn at school of arithmetic: addition, multiplication etc. But the basis of the redeeming arithmetic is a movement that distinguishes *one* from *zero*, something from nothing.

The path from Greece and the east was lost when identifying the *one* and *zero* separately as symbols 0 and 1. In the redeeming arithmetic, *zero* and *one* are represented only by the meaning that their relationship realises. I am an identity only in so far as I confront the *zero* of my non-existence. The sum total of all such journeys of *zero* to *one*, of non-existence to existence, of potential to expression, is also the elucidation of the meaning of *one* generically as a statement over nothing.

This is what Pirsig meant earlier in his description of quality. Quality is the meaning of *one*, the Godhead of the Unity, as the generative source of all the action. Spirit and God are no longer isolated from the system of science. Unity has a meaning as existence over emptiness, as the sum total of all the ways in which individuals, ideas and systems uniquely and creatively find their own identity in a test of nothing. That is the point of the new arithmetic. The journeys dark to light, *zero* to *one*, potential to expression, add up to a unique calling of *one* as quality over emptiness.

This is a complete turnaround. Instead of having to lock my identity away in protection of an isolated integrity, my whole journey in life speaks itself uniquely into the quality of unity as it declares itself over emptiness. In this quality everything is included: art, science, my failings, my wrong turns, difficult relationships, travelling . relating *zero* and *one* in myself calls forth the quality of *one* as it knows itself as meaning over *zero*.

Self-refuting logic

We have a new notion of logic. The very connotation of a logic of existence leads to its own refutation. The openness of emptiness, leads to synchronous form. Logic is about the refutation of any finite assumption into its opposite. Logic is by its own reasoning not the first element of existence.

Logic is a supposition that considering all the evidence, necessarily leads to its negation as the only possible truth. Then this opposite becomes the new supposition and in weighing all the evidence for the negation, one is forced logically to return to the original supposition etc. The potential for form, relates equally to nothing as it does to existence.

Thus instead of deducing step by step, as we like to do, this is space, this is time, this is matter, we allow the suggestion of non-space to lead to the development of an arrival at a paradoxical meaning, asserting existence, even though this supposition at once negates itself.

Narrowing down the field of logic somewhat, from a theory of everything, logic goes no further than denying its own supposition in consideration of the evidence and formulating its opposite tenet. The logic of the particle is the wave and of the wave the particle, not as an aberration, but as the very practice of physics. The understanding of light as the darkness, and the darkness as the light, is the furthest revelation logic can bring to us as thinking beings.

Nowhere is this better illustrated than by the story of logic in science itself.

Intellectual track – materialism

From 1850 onwards, science attempted to establish on secure fundamentals a complete logic based on reason. For instance in 1854 George Boole isolated the elements of logical thinking in his pivotal work, *The Laws of Thought*. In 1859 Darwin published *The Origin of Species*, in which again all higher motives in life were reduced to a lowly process of inheritance of acquired traits through natural selection. And in the early 1860s Maxwell overturned the order of physics with the mathematical derivation of the laws of electromagnetism, including the nature of light.

Nineteenth-century mathematics imputed that actually everything, including infinity was graspable within a single comprehensible theory. Everything conceivable shared a single mathematical ground of derivation. Everything that could happen was linked by a mathematically rigorous process of deduction of one thing from another.

Hilbert and Russell in the early twentieth century saw for mathematics the goal of arriving at the ultimate form of logic. The aim was to distil everything of mathematics to its final form of deduction out of its own principles. But Gödel in 1931 showed in his *Incompleteness Theorem*, mathematics itself as a logical endeavour could never be seen as a coherent static picture of a defined reality.

Had science defeated its own purpose?

Yet this self-negation of its own proposition allowed for a second round of understanding, based on the premise of the paradoxical. In allowing the paradoxical to hold the negating argument of Gödel, the inner circle of thought was re-legitimised in the wider circle of understanding brought by Spencer Brown and Kauffman building on the philosophy of Kierkegaard, Goethe and Sartre, as illustrated in *Figure 20*.

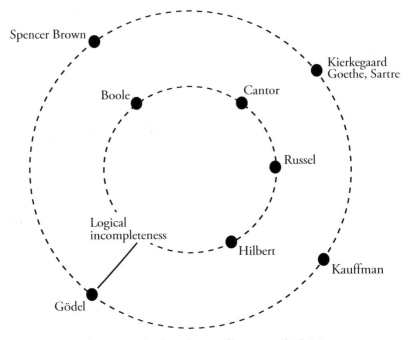

Form disproving the foundation of logic on which it is set

Figure 20. Double circle of the negation and assertion of a living logic.

Boole: isolating truth from falsity

Early science had understood the finite world as the expression of God's justice and balance. The planets, the earthly bodies, all moved according to the perfection of realisable law. Man lived in the perfection of a cosmos whose rhythms he could fathom. Since Newton, science had seen itself as representing through subtle concepts, the working of the universe, the turning of planets, the motion of bodies.

The 1850s saw a big shift in this relation. This upheaval centred on the French and other revolutions in Europe and the return to a popular understanding. There was a move against elitism and towards the accessible depiction of events. In art there was the movement of realism aimed at presenting objective reality without embellishment. Darwin in biology, Maxwell in physics and Boole in logic, reclaimed from outside authority the independence of scientific reason. The random chance of evolution, the fields of electromagnetism, the logic of true and false, gave back to thought, dominion over the world.

For instance the title of George Boole's 1854 book is itself telling: *An Investigation into the Laws of Thought on Which are Founded the Mathematical Theories of Logic and Probabilities.* Boole's work often known as *Laws of Thought* begins:

> The design of the following treatise is to investigate the fundamental laws of those operations of the mind by which reasoning is performed to give expression to them in symbolic language of a Calculus, and upon this foundation to establish the science of Logic and construct its method. (Boole 1854)

In Boole's work one already feels the first reverberation of the challenge to the skies, the heavens of old, to give up existence to the pure logic of the mind. Mathematics is subtly repositioned to place reasoning itself as the foundation of science. Science is given the confidence of defining itself outside external authority as to the nature of its endeavour.

The universality now moves to the process by which thought is able to assemble the myriad elements of existence into an abstractly conceived order. Boole's work is a new type of move in science. Boole uses mathematics not to describe existence but to claim a ground (the pure logic of the mind) as the foundation of science.

Boole's work sets a test to finite existence. Is it by the logic of reasoning alone that the universe is able to reveal the nature of its own form?

Following Boole, there were various attempts to define exactly the complete representational foundation on which a finite world could be built, including Cantor's set theory.

Cantor: an infinite challenge

Georg Cantor, who was born in 1854, significantly the year of the publication of Boole's *Laws of Thought*, devoted his life to the internalisation of the infinite from the heavens into the argument of the number system itself. Cantor often became depressed by the irrationality of this endeavour to try to fit into number what was by its nature ungraspable. Nevertheless it was a struggle he pursued over his whole life.

The way Cantor came to the relation to the infinite was through sets, collections of objects and laws similar to Boole's, which connected sets together. Since the challenge from Boole's time was to see in mathematics the pure logic of the mind, then there was now no boundary upon what mathematics could attempt by way of demonstrating the nature of existence. Thus for Cantor it became a theological issue, for which he felt called by God, to demonstrate through number the infinite basis of the ground of existence.

The discovery of a procedure by which one could move from infinite sets to more numerous infinite sets in ordered steps, indicated to Cantor that the infinite was not separate from the finite. The finite completed itself through a real relation to the infinite.

Cantor's attempt was to allow the infinite a place within the finite. Cantor used symbols of *aleph*, the first letter of the Hebrew alphabet, with a subscript for designating the various degrees of infinity one could reach in extending one infinite set by a simple procedure into a larger infinite set by concocting original elements out of the existing members of the set. These symbols are still in use today.

If the finite has a structure then Cantor showed that so does the infinite.

This aroused much criticism from both scientists and theologians. Hermann Weyl wrote:

> Classical logic was abstracted from the mathematics of finite sets and their subsets ... Forgetful of this limited origin, one afterwards mistook that logic for something above and prior to all mathematics,

and finally applied it, without justification, to the mathematics of infinite sets. This is the Fall and original sin of [Cantor's] set theory. (Weyl 1946)

The perception was created that Cantor had included the infinite within the finite system of mathematical logic. The idea of Cantor was further brought into question by a simple paradox of the young Bertrand Russell in 1902, which showed how sets could be defined that were inconsistent with their definition of membership. This brought into question the notion of sets as being a fundamental logic.

Cantor's life, dramatic as it was, with discoveries punctuated by long periods of depression partly about being misunderstood, marks a change of era to a belief in the foundation of rationalism. This shift of emphasis from a relation to the infinite, to a system extended to include the infinite, was subtle and pervasive.

Finally science asked the question about what was the ultimate set of symbolic representations, called mathematics, that gave a coherent statement about the world from their own internal deductions.

Nothing but mathematics

The effort to use the pure logic of the mind as the gateway of realising a physics of the universe, without outside dependence, led Whitehead, Russell and others to explore the *Principia Mathematica*, the very ground rules of logic on which the whole edifice of mathematics could be securely founded. The work looked to definitively set down the relation between the axioms that are the finite statements with which one begins and the set of statements that mathematics allows one to make about reality on that basis. The impulse behind the work was to establish the ground for the growing reliance put on mathematics.

As Weyl wrote in his essay 'Levels of Infinity' in 1931:

But we can hardly still believe today that something comprehensible is to be found behind these theories of Cantor. Thanks to the critique of H. Poincaré, B. Russell, Brouwer and Skolem, and others, we have gradually had our eyes opened to the untenable logical positions from which the method of set theory proceeds.

A rescue of mathematics without diminution of its classical integrity is only possible, as D. Hilbert first recognised, through a radically changed interpretation of its meaning, namely through a formalisation in which, in principle, it changes from a system of knowledge gained through insight into a game carried out to fixed rules, with signs and formulas. Every mathematical statement is transformed into a meaningless formula built out of signs, and mathematics itself transforms into a game regulated by certain conventions, quite compatible to chess. One or several formula count as axioms ... formal rules of inference reign, according to which new formula can be produced, i.e. 'deduced' out of previous formulas.

The mathematician seeks a skilful linking of moves into a final formula that will let him win the proof-game by proper play. Up to this point, everything is a game, not knowledge; however now the game is made into the object of knowledge in '*metamathematics*,' as Hilbert puts it: it should be acknowledged that a contradiction can never appear as the final formula of a proof. Hilbert wants to establish with certainty only this *freedom from contradiction*, not the truth of the content of the analysis. (Weyl 1931, p. 26–27)

Gödel's quality beyond logic

Gödel, in 1931, famously sunk this attempt by showing in his *Incompleteness Theorem* that no finite system of deduction could ever be complete! No system of finite statements could ever meet this aim by the very limited nature of mathematics itself. There would always be a statement that transcended the finite system's ability to establish its own consistency.

Using an argument much like Cantor's for the unlimited nature of any finite collection of elements, Gödel showed in a proof, that there would always be statements that were undecidable whether they be true or false on the basis of the axioms. Gödel showed that mathematics was necessarily incomplete as an internal system of inference.

Gödel dismissed the very attempt of mathematics to stand in its own finite ground of hypothesis and deduction, alone, without outside dependence. This was quite an earth-shattering event. It questioned the central assumption that reasoning was the only coherent way to establish a view of the universe. Mathematical reasoning on its own, if not combined with some living test of its validity, was by its own nature incomplete.

To argue one finite step after another finite step could in principle, by Gödel's proof, never arrive at a complete statement of existence.

In 1956 Russell concluded:

I wanted certainty in the kind of way in which people want religious faith. I thought that certainty is more likely to be found in mathematics than anywhere ... But after some twenty years of arduous toil, I came to the conclusion that there was nothing more that I could do in the way of making mathematical knowledge indubitable. (Russell 1956)

Spencer Brown

While many saw the *Incompleteness Theorem* as a void into which man had fallen off the edge of the validity of reason, Spencer Brown took this as a challenge to embrace uncertainty and paradox in a novel logic. Spencer Brown challenged the attempt to base logic on static entities of existence, which at the same time didn't exist.

In the whole science of physics there is no such thing as a thing. Hundreds of years ago we carefully forgot this fact, and now it seems astonishing even to begin to remember it again. *We* draw the boundaries, *we* shuffle the cards, *we* make the distinctions. In physics, yes physics, super-objective physics, solid reliable four-square dam-buster physics, clean wholesome outdoor fresh-air family-entertainment science fiction superman physics, they don't even exist. It's all in the mind.

If you separate off this bit here (you can't really, of course) and call it a particle (that's only a name, of course, it's not really like that, more like waves really), only not really like that either, not really, space is not what you think, more a sort of mathematical invention, and just as real, or just as unreal, as the particle. In fact the particle and the space are the same thing really (except that we shouldn't really say 'thing'), the sort of hypothetical space got knotted up a bit somewhere, we don't know exactly where because we can't see it, we can only see where it was before we saw it, if you see what I mean, I mean even that's not what it was really *like*, it was waves (or rather photons) of light carrying a message that may well be very unlike the thing, sorry, particle (remember this is only an *abstraction*, so that we can talk about it (it? sorry, we don't have an *it* in physics)) where it (sorry!) came from.

The significance of this way of talking, which, as everybody knows, is called modern science, is maintained by means of a huge and very powerful magic spell cast on everybody to put us all to sleep for a hundred years, like that nice Miss Sleeping Beauty, while the amusements are being rigged up. (Spencer Brown 1974)

For Spencer Brown two things are true: (1) what holds in mathematics are not things but distinctions; (2) the medium of distinction is paradox, where many things that may become true affect each other. This applies to mathematics as a whole. What is to come out of the huge effort of mathematics is a distinction of the spirit as a living quality, the bizarre ground on which all the inconsistencies of science may resolve.

The set of finite statements about existence is incomplete. Given that the world is logically consistent there must be a way of retrieving from this inconsistency, a further round of logic that recovers from finite incompleteness, a whole statement of reality. A second circle of logic compensates for the break between finite assertion and realisation by travelling from paradox to form. This second circle of logic allows the meaning itself to realise the coherence of the elements that are thereby signified in relation. It is the organic, whole completion of the causal relation between finite particulars.

The title of Spencer Brown's book (*Laws of Form, 1969*), deliberately contrasted with Boole's *Laws of Thought*. Instead of logic being in the operation of the distinguishing of true and false, as these are discerned by the mind of the logician, the essential logical element is the circumscribing of emptiness into the form of existence. While Boole (and in his footsteps Russell) shuns paradox, where everything is either true or false, Spencer Brown introduces paradox as the essence of the exercise.

The statement logic can make about existence, is limited by its juxtaposition with emptiness.

Falling paradigm

This challenge to the foundation of science (from the exact certainty of reason to the vagaries of the vulnerability of chance), is illustrated in the attempt by Spencer Brown to have his logic validated by Bertrand Russell. Spencer Brown tells the story of his meeting with Bertrand Russell, in an attempt to have Russell endorse the manuscript.

Russell invited Spencer Brown to give him a week long seminar on his work, but by the last Friday afternoon, they had made no progress in making a bridge between their opposing viewpoints.

> 'And how do you deal with the propositional functions?' he [Russell] snapped. 'Bertie, I don't,' I replied. 'Oh' he said. 'And why not?' 'Because I'm not a good enough mathematician,' I said. 'Oh,' he said. His manner changed, and he grew confidential. 'You know, I have never admitted this to anyone before, but I, too am a very bad mathematician. Come, let me see what you have written.'

Russell was then able to appreciate the system of logic Spencer Brown had created:

'You have made a new calculus, of great power and simplicity, and I congratulate you!'

Whether actual or symbolic the discovery was celebrated with whisky, resulting in Russell becoming unsteady and holding on to a huge spiral bookshelf for support.

As he grabbed it, there came an ominous crack of splintering wood from the pedestal at the base, and the whole edifice tottered crazily towards him, showering down its contents. And as each section of shelves, relieved of the burden of its books, became lighter, the heavier sections, still full of books, revolved towards Bertie and showered their contents over him again. I watched fascinated … until the entire bookcase was empty and Bertie had disappeared under a pyramid of books about five feet high.

The noise and the shock, while it lasted, was like an earthquake, and it brought Edith hurrying in from another part of house to see what had happened. … Together we rapidly dug Bertie out, stood him up, and dusted him down. He turned to Edith and said, 'It's perfectly all right. There's no need to look worried. We were just celebrating what clever fellows we are!' (Spencer Brown 2009, p. 114–16)

There was something symbolic in this towering edifice of books falling over Russell as a new logic was revealed. The attempt to found science on a self-contained static platform had ended. Spencer Brown's work marked a move to a different era of logic. Mathematics saw through Boole's attempt to derive a calculus of the elements of thought needed to sort the world into true and false propositions. The meaning of true in relation to false was established only in the test that challenged existence with emptiness in paradoxical juxtaposition.

Wheel of physics

The question that physics had been arguing since Einstein and Bohr is articulated by Arthur Fine particularly poignantly.

> Bohr thus views the product of conceptual refinement as a wheel like structure: a central hub from which a number of disjoint spokes extend. Different explorers can move out separately along different spokes, but, according to Bohr, the reports they send back will not enable one to piece together an account of some region between the spokes or of a rim that connects them.
>
> Einstein's dispute with Bohr (and others) is a dispute over this wheel like structure. Einstein asks whether the spokes must really be disconnected, could there not at least be a rim? This is the question as to whether the quantum theory allows for a realist interpretation. … And Einstein asks whether the spokes must really be made of the same material as the hub? Must we, that is, stick with just the classical concepts? (Fine, p. 21)

Bohr's rim is purely made of philosophical principles that he basically invents, without further justification, to preserve the consistency of the inner ring of physics. *Complementarity, uncertainty principle, correspondence principle* and probability interpretation of concepts, are all things added into physics as philosophical asides, in order to save the inner circle of conceptual consistence. Bohr's world is one of emptiness. Under this view, reality is a conceptual illusion held together by physics and philosophy.

In contrast Einstein is seeking a realist picture. Can we go beyond the conceptual structure of the inner hub and imagine a different kind of outer rim? Einstein's idea is then more about the *one*, as the reality of things as they exist in the world.

Einstein wrote to Schrödinger:

> Your claim that the concepts momentum and position will have to
> be given up, if they can only claim such 'shaky' meaning seems to
> me to be fully justified (Einstein 1928, in Fine p. 18)

Einstein is for giving up the conceptual understanding, rather than
keeping hold of it at whatever cost. Fine's analogy of the nature of
the outer rim fits perfectly with the paradoxical circle of Spencer
Brown's mathematics.

Spencer Brown's mathematics describe a circle in which context,
identity and time, are woven together in a self-developing whole,
a structure that maintains itself by its own self-production. It is
thus possible to see the analogy of the rim and the hub in this way.
The paradoxical rim communicates with the conceptual hub in
the different ways that multiple perspectives of potential can speak
themselves into distinct meanings.

As Fine so eloquently puts it, the rim that quantum theory
needs is of another material than the hub of precise physics. To
understand the rim one has to let go the certainty of the centre,
the *one* of existence. The rim does not need the precise linear
logic of construction of the hub. The job of the rim is to roll a
universal impulse around the circle of the paradoxical tension. For
Spencer Brown, this universal impulse may have nothing to do with
mathematics at all. The only quality required of it is to realise a
cycle through the paradoxical elements to its own origin.

In fact, all we have to do, as Goethe proposes, is have a science
based on the ability of our seeing to become one with the world.
This is the only test that our participation needs, in order to travel
the outer rim of the wheel of physics. Barfield describes this:

> Participation as an actual experience is only to be won today
> by special exertion: that it is a matter, not of theorising, but of
> imagination in the genial or creative sense. A systematic approach
> to final participation may therefore be expected to be an attempt
> to use imagination systematically. This was the foundation of
> Goethe's scientific work. In his book on the *Metamorphosis of
> Plants* and the associated writings descriptive of his method, as

well as in the rest of his scientific work, there is the germ of a systematic investigation of phenomena by way of participation. The processes [are] grasped directly and not, as hitherto since the scientific revolution, hypotheses *inferred from* actual phenomena. (Barfield, p. 158–59)

The concepts at the heart of the debate between Bohr and Einstein are not there *a priori* as the foundation of the physical system. The spokes of this wheel are the paradoxes one encounters which have the aim of going beyond the finite description of the world. In the logic of the paradoxical, something whole and creative is allowed to enter. To understand the conceptual hub, one must travel along the rim of the vulnerability to emptiness in making meaning, as illustrated in *Figure 21*.

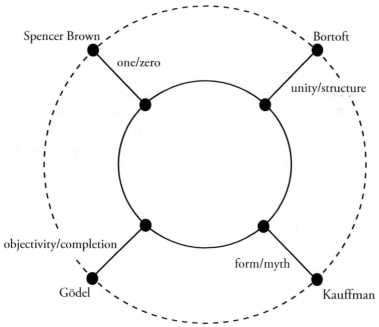

The rim of paradox circumscribing complementary aspects of physics

Figure 21. The wheel of physics.

The overlay of potentials contains a paradoxical judgment which nevertheless contributes to a meaningful resolution over the set of all such potentials and their forms. Particularly of significance is the formulation of a self-denying existence. For Spencer Brown this involved showing that if one begins with the paradox of an existence that negates itself, one could still arrive at meaningful form. Starting with nothing the statement realises existence, which iterated again turns back to nothing and then to an existence, and so on. Or starting with existence it produces nothing, and then an existence, and so on.

What this paradoxical statement is saying, is that emptiness is to be taken as the birth of form. The way to understand the paradox is to see the oscillation of emptiness and form as the basis of an unfolding over time. So emptiness is the mother of a form that then dies back to an emptiness that is creative once more of a form, as a periodic foundation to time. One then arrives at the two basic statements about existence:

> Emptiness is the nature v. existence (western)
> The nature of existence is emptiness (eastern)

It is then valid to start an investigation in the paradox of mathematics by considering the philosophy of meaning. For, the outer rim of the wheel of physics is the movement that holds whole meaning over the cycle of experiences and understandings in which life is paradoxically implicated.

Kierkegaard expresses most clearly in philosophy, how paradox becomes the gateway to resolution. He describes existentially in the life of the individual, the tension between the finite synthesis (the inner circle) and the infinite challenge of existence (outer circle).

The test of the infinite – Kierkegaard

In 1849, Kierkegaard, on the edge of the halfway threshold of the nineteenth century where the infinite was still untamed, began *The Sickness unto Death* with the following:

A human being is a synthesis of the infinite and the finite, of the temporal and the eternal, of freedom and necessity. In short a synthesis. A synthesis is a relation between two terms. Looked at in this way a human being is not yet a self.

Kierkegaard describes the inner circle as 'not yet a self.'

In a relation between two things the relation is the third term in the form of a negative unity, and the two relate to the relation, and in the relation to that relation.

Out of the paradox of the inner circle of human aspiration to selfhood, is formed the outer circle of the negative unity.

Such a relation, which relates to itself, a self, must either have established itself or been established by something else.

Because we are in the finite and always aspiring to the infinite, we are cast into this world of separation even though we are longing for unity. Kierkegaard characterises this state as despair. He then speaks of how to go beyond this state.

This then is the formula which describes the state of the self when despair is completely eradicated: in relating to itself and in wanting to be itself, the self is grounded transparently in the power that established it. (Kierkegaard, p. 43–44)

The division between self as identity and self as immersive participation, the synthesis of two things, a tension between poles of opposition, allows for a power that establishes the ground of the two things. The establishing power is external to the system of self. The establishing power of relationship doesn't come from within the self, from the two poles of synthesis, it comes from outside. In facing up to the synthesis, one has the opportunity to discover what is beyond that synthesis.

How can we talk about the two different realities without getting lost in the separation of them? For Kierkegaard the way is to allow both to exist and to allow the poles of opposition to journey in their contradiction, to a resolution that may be entered anew as the birth of logic. In confronting the scientific world and the experiential world, you allow this third aspect of relation that establishes how the opposition reveals its whole quality.

Science has come across paradox in quantum theory, but has tried to keep the terms apart: the wave and the particle, the momentum and the position. We can never measure the two at the same time. So in meeting paradox within physics, we have tried to keep the different aspects apart and say there is separation. But there is another way of dealing with this synthesis, between two things that are apparently opposite and contradictory. In allowing this process where the third unifying term is the power from outside that establishes the relation of the synthesis, we find a whole different way of seeing physics.

In the meeting of opposites, instead of the world standing in complete deadlock between two voices, the two attitudes are the two poles of a synthesis. The more we go down into separation, the more the two sides are entrenched. In holding the two attitudes as not yet formed as a self, as not yet existent, as not yet being, there is this third quality that can come in. We can turn science on its head, for instead of getting stuck in what is the reality of separation, we use this synthesis of opposites to allow something to reveal itself, with a profundity that deepens with the extreme nature of the opposite things, brought into relation.

Phenomenology completely turns upside down the premise of western science, that we are living in a synthesis, as subject and object. Upon that very ground, the world falls apart in a completely revelatory and unique way.

Goethe's affirmation

For Goethe, dark and light are brought into relation through colour. Instead of seeing dark and light as separate things, we see colour as a phenomenon that establishes the relation of darkness and light. Dark and light are the synthesis of opposite poles and the colour establishes the relation between the two.

In Goethe's *Faust*, the main character surrenders to separation as an emptiness, that thereby he might know the other pole of salvation better. Faust enlists the help of Mephistopheles to fall into separation without any reprieve from its twisted outcomes. He experiences the consequences of separation, multiplied time upon time, in the attempt to meet desires and curiosity with new acts of trickery, to endlessly immerse himself in a world that is nothing but the play of matter, devoid of any meaning, taken to ridiculous and tragic extremes. As in Goethe's colour theory, the darkness is a necessary contrast in which to experience the light. The light can never exist on its own as one emporium, separated from the dark on the other.

Goethe shows us that the goal of life is beyond the particular tangles in which our adventures in the world of separation have entrapped us. Life is about overcoming the separation, by the actions that reveal to us the ground of the double nature of being – in my case, as a computer programmer and a seeker after meaning. Our understanding of everything – the earth, the fields, humankind, shares this double nature and is open for renewal, taking us out of the separation of abstraction from participation. Existence seeks together the action in which separation is relativised, the ground of dichotomy, between God and Mephistopheles, revealed through our own experience.

Our engagement with the world is able to reveal to us the power that establishes the ground of our being. Our engagement in the world actually reveals the ground of who we are underneath our acting in that way.

The opposites are not yet a self and yet we need the opposition for the possibility of self to be there. We need opposition to feel the potential for self and in finding the power that establishes us, we become a self. But it is a transient thing. We need the separation to have that act of finding our self. So by nature we are always finding ourselves to lose ourselves, we are always in that despair. We explore the dichotomy of whole and part, we create a synthesis of opposites of part and whole.

Goethe's method is that once we have created the dichotomy between whole and part, we seek the ground that establishes the dichotomy that it is both a part and an archetypal quality. In allowing those opposites, the power that establishes the opposites can reveal itself as a movement that establishes both the parts in their existence and the whole, the identity. In allowing the opposite to exist, we can experience the power, the movement that is underneath those two things.

The synthesis may end in conflict, which is then finite because it is separate, or move to order that is infinite. Before us is either the dissolution by separation into an endless recrimination and stalemate of views, or the overcoming of separation into an order that reveals the infinite. This is then the singular challenge of all existence, whether our own movement is to destruction, extending the separation, or to creation, investing the separation with new meaning.

The opposition of order and freedom can be resolved into order or blow apart into conflict. The very aim of science since the 1850s has been to incorporate the infinite into the finite, to try to conceive of the world of separation as something complete in itself. This in a Faustian way, forces us to reconsider the relation of ourselves to meaning.

Sartre's tune through nihilism

I was given the book *Nausea* written by Jean-Paul Sartre. *Nausea's* main character and narrator Roquentin, is completely alienated from the world and its meaninglessness. He is sick of the world and its existence. Sartre drags the reader through this excruciating nothingness, page after page. At the end, Roquentin is in a café. He is looking into his beer glass, experiencing how meaningless it all is. And then the record plays on the turntable:

> And I too have wanted to *be*. Indeed I have never wanted anything else; that's what lay at the bottom of my life: behind all these attempts which seemed unconnected, I find the same desire: to drive existence

out of me, to empty the moments of their fat, to wring them, to dry them, to purify myself, to harden myself, to produce in short the sharp, precise sound of a saxophone note. That could even serve as a fable: there was a poor fellow who had got into the wrong world. He existed, like other people, in the world of municipal parks, of *bistros*, of ports and he wanted to convince himself that he was living somewhere else, behind the canvas of paintings, behind the pages of books, behind gramophone records, with the long dry laments of jazz music. And then, after making a complete fool of himself, he understood, he opened his eyes, he saw that there had been a mistake: he was in a *bistro*, in fact, in front of a glass of warm beer. He sat there on the bench, utterly depressed; he thought: I am just a fool. And at that very moment, on the other side of existence, in that other world which you can see from a distance, but without ever approaching it, a little melody started dancing, started singing: 'You must be like me; you must suffer in strict time.' The voice sings:

> *Some of these days*
> *You'll miss me honey*

And there is something that wrings the heart: it is that the melody is absolutely untouched by this little stuttering of the needle on the record. The record is getting scratched and worn, the singer may be dead; I myself am going to leave, I am going to catch my train. But behind the existence which falls from one present to the next without a past, without a future, behind these sounds which decompose from day to day, peels away and slips towards death, the melody stays the same, young and firm, like a pitiless witness. (Sartre, p. 248–49)

Roquentin finds salvation in this song, this completely ridiculous song which somehow straddles his non-existence and that pure being he wants to be. He jumps out of the separation he has got himself in: between the pure being he wants to be and that meaningless world of warm beer and bistros and waiters. He realises that it is in our very nature to overcome that separation. We can become a real self when we allow that creative energy to come in.

Experience is the primary thing. The world isn't going to be saved by creating more separation. It is in everyone's experience to find their own way of crossing the synthesis, the opposites. There is no guidebook telling us how to do it. To notice and understand how our energies are working through attention is a good start. A composer wrote that song, which moves Roquentin all those years later. Happening is a primary medium and happening has within it everything, the beauty, the science, the necessity, the freedom, the temporal and the eternal. All beings make mistakes, all beings are the subject of a synthesis but all beings have the capacity to return that synthesis back to a ground.

Renewing relation

As odd as it is to talk of the limitation of logic, we trust implicitly in the quality by which the world speaks itself in our own lives. Often when we are harassed, we are unable to escape from the details of a problem and we move around its finite elements in an ever darker hole of confusion.

Then we accept that something out of our own hands is involved in the whole enterprise. We surrender. We take a breath and allow the situation to come to a meaning by itself. This inclusion of freedom from our own finite meddling in the situation, immediately brings new perspectives. We are able to imagine new ways of approaching the problem, that have intrinsic worth, rather than the random finite activity we were engaged with. The way the world speaks itself is not something far off and impossible to imagine, it is actually close at hand.

This facility to ride the wave between existence and non-existence as a way to hear the world in its statement is one at which scientists are adept. The 'aha!' moment of thinkers arises from nowhere and remakes a path through all the fragments of unfolding. Poincaré describes such an experience as follows:

Just at this time, I left Caen, where I was living, to go on a geological excursion under the auspices of the School of Mines. The incidents of the travel made me forget my mathematical work. Having reached Coutances, we entered an omnibus to go to some place or other. At the moment when I put my foot on the step, the idea came to me, without anything in my former thoughts seeming to have paved the way for it, that the transformations I had used to define the Fuchsian functions were identical to those of non-Euclidean geometry. I did not verify this idea; I should not have had time, as, upon taking my seat in the omnibus, I went on with a conversation already commenced, but I felt a perfect certainty. On my return to Caen, for consciousness's sake, I verified the result at my leisure. (Hadamard, p. 13)

The *one* of existence tells its meaning to the *zero* – the lost endeavour of Poincaré to reason through the mathematical problem by thought alone. In forgetting his mathematical work, in coming to rest in the *zero* of unformed potential, the *one* then stated itself as a matter of course.

Einstein also made evident the physical movement in which his intuition became grounded in theory:

The words or the language, as they are written or spoken do not seem to play any role in my mechanism of thought. The psychical entities which seem to serve as elements of thought are certain signs and more or less clear images which can be 'voluntarily' reproduced and combined.

There is, of course, a certain connection between those elements and relevant logical concepts. It is also clear that the desire to arrive finally at logically connected concepts is the emotional basis of this rather vague play with the above mentioned elements. But taken from a psychological viewpoint, this combinatorial play seems to be the essential feature in productive thought – before there is any connection with logical construction in words or other kinds of signs which can be communicated to others. (Hadamard, p. 142)

Poincaré offers this insight in the creative process, likening atoms to ideas.

> We think we have done no good, because we have moved these elements a thousand different ways to assemble them and have found no satisfactory aggregate. After this shaking-up imposed by our will, these atoms do not return to their primitive rest. They freely continue their dance. The mobilized atoms undergo impacts, which make them enter into combinations among themselves or with other atoms, at rest, which they struck against in their course. (Hadamard, p. 47)

Poincaré uses the notion of atoms (before much was even known about the physical atoms of quantum theory) to be suppositions of our thinking that engage in a play to resolve the proposition of existence as a meaning out of emptiness. The standing of initial transient ideas between *zero* and *one*, play and realisation, including and excluding, surrender to the quest of finding the meaning that gives a quality of connection to all these previously disassociated ideas.

The paradoxical entities of atoms hold together, in the context of their association, a space for some larger unifying idea to come in and transform what was already there, neatly into something new. The atom concentrates the play of the outer rim of renewal around the inner hub of structure. The atom is the unity of moving rim and structured hub that allows whole transformation to occur through finitely limited parts. The meaning that connects all the concepts, transforms the nature of the path, in which the hub is logically conceived.

In other words, the *Incompleteness Theory* of Gödel is replaced by an *Existential Completion Cycle!* Theories do not have to extend themselves unbounded towards infinity. They only have to hold the cycle of transformation by which an old unity moves round wholly to a new one.

There is no difference between the quality of the artist and the scientist or the endeavour of a materialist and a mystic. The more diverse are the elements, the greater and more original is the sounding of unity through emptiness.

Newton, no less than Leibniz, worked into the very depth of the phenomena of motion but whereas Newton held this search into nature on finite conceptual terms, Leibniz went through experience into the quality of what existed infinitely for itself. The scientist and the mystic are involved in the same dynamic of penetrating into the nature of things, only the other way around from each other. The scientist engages with the infinite. He has the patience to embark on a many year course of study, with the finite being the vessel of conceptual relation into which nature may be revealed. For the mystic, in contrast, the finite is the instrument surrendered as the medium in which the infinite note of nature may sound in timeless spirit.

The infinite is thus the key that helps us know ourselves as active existences out in the world, both individually and collectively engaged. What is the infinite? The infinite is any domain of existence outside the finite realm of our own busy-ness. The infinite is that horizon of reference that is able to remain itself, outside of our work from potential to action. In other words we know the infinite as the meaning that holds true to our worth beyond the investment of our work.

Language

Another way we see this cycle of meaning is in language. On the one hand, language is a precise structure of words and grammar that every student has to learn. But if that was all there was to language, speaking would be very tedious. Around this hub of structure, language is also the moving rim of the making of meaning. As example, when I was addressing a class on what was the nature of the space of possibilities, language itself came to my aid.

I speak. One way this can happen is to look in the dictionary and work out all the words that I want to use and then try to assess analytically how to put these words together to convey a meaning. Another way, closer to what actually happens when I speak, is that I enter into a possibility space. I have a feeling of what I am trying to communicate. I know the potentials of these words I want to use to communicate. I let go of thinking I can control what is happening. I let the words in their possibility suggest their own way in which they are going to convey a meaning. I don't first fix things and then try to put a meaning in them. I jump into the possibility space and let the words be the mediators of the meaning I am trying to express. That possibility has the quality of navigating to something that isn't yet clear when I begin. So when I begin I am not yet sure how I am going to say what I am going to say. But in entering that possibility space, jumping off the end of knowing what I am going to say, the potentials enter the possibility space and arrange themselves to communicate the meaning.

I let that possibility space travel through my attempt to communicate, to the point it delivers meaning. Because we are only dealing with possibility, the possibility space resolves the potentials in something that lies ahead when I begin. I meet with reality somewhere ahead in time, where a destination resolves the possibility.

In language, time is not captured in its system. But the way words lend themselves to the formation of a meaning, holds within it the ordering of time. Past and future are distilled into a statement of reality that charges the finite system of time to come into line with its dictum. This is the true power of the word. That it is literally beyond the limits of the finite way in which we divide reality in our internal division of what is what. The word spoken right changes the world.

A clearing appears. What we are looking at is a possibility space, a clearing that we each feel in our own way. Our journey is about arriving at a space that we all experience. That moment where we touch something together, is a light-filled moment where something we are seeking discloses itself.

This notion of dynamic language brings us back to a statement of Bohr:

> The elucidation of the paradoxes of atomic physics has disclosed the fact that the unavoidable interaction between the objects and the measuring instruments sets an absolute limit to the possibility of speaking of a behaviour of atomic objects which is independent of the means of observation.

We are here faced with an epistemological problem quite new in natural philosophy, where all description of experience has so far been based on the assumption, already inherent in ordinary conventions of language, that it is possible to distinguish sharply between the behaviour of objects and the means of observation. This assumption is not only fully justified by all everyday experience but even constitutes the whole basis of classical physics... (Bohr 1938, p. 25–26)

Bohr firstly applies language to the job of description of the behaviour of atomic objects and secondly rules out language as capable of rendering the subtlety of object-observation interaction. He takes the imprecision of language as a warning that we must jettison our own sensibility and rely only on the mathematics. In holding to the exactness of concepts, he has to excuse the imprecision, by inventing a purely philosophical encompassing of measurement.

In this respect, we must, on the one hand, realize that the aim of every physical experiment – to gain knowledge under reproducible and communicable conditions – leaves us no choice but to use everyday concepts, perhaps refined by the terminology of classical physics, not only in all accounts of the construction and manipulation of the measuring instruments but also in the description of the actual experimental results. (Bohr 1938, p. 26)

But language is exactly the metaphor we need to describe the rim of the making of word-conceptual meaning. Language is adept at travelling around the rim of inexactness, allowing the concepts of words to form as appropriate to the meaning of their whole cycle. The looseness of the atomic structure is to allow many possible conceptual forms to be spoken together into a viable meaning. The language of living form that we see about us, all follows from the lucidity in which the paradoxical cycle turns a whole meaning about the hub of a conceptual centre. Spencer Brown's logic of paradox allows multiple conceptual dimensions of ambiguous possibility to reveal a whole meaning.

Quaternions

Lou Kauffman, Professor of Mathematics in Chicago, takes further Spencer Brown's paradoxical logic that argues science backwards from its conclusions to the statement of its definitions. The advanced method of working, finds in context, the conditions for the elements of its logic to be given meaning.

Kauffman illustrates the paradoxical logic in the quaternion scheme:

> A remarkable symbolic scheme, devised in the 1800s by the mathematician William Rowan Hamilton, gives the pattern underlying these compositions of rotations. The scheme is an algebra associated with i, j and k that Hamilton called the *quaternions*. (Kauffman 1982, p. 127)

The quaternions are a number system of triple imaginary elements. Hamilton's insight came in 1843 crossing Brougham Bridge in Dublin. With no writing material at hand, he carved the magical interdependence relationships in the stonework, (commemorated to this day by a Quaternion Plaque). The next day he wrote in a letter:

> And here there dawned on me the notion that we must admit, in some sense, a fourth dimension of space for the purpose of calculating with triples ... An electric circuit seemed to close, and a spark flashed forth. (Hamilton, p. 489–95)

Let us take for instance the three paradoxes of the first three chapters and identify them with Hamilton's elements.

i = [the dimensions of] **division/unity**
j = [the dimensions of] **darkness/light**
k = [the dimensions of] **past/future**

Providentially for our chapter heading, quaternions are the particular mathematics that describes the make-up of the rotations of a three-dimensional cube, as a dice (Kauffman 1982, p. 125–34). It is through the paradoxical properties that dice so ably provide the foundation of delivering chance events. For the logic of the dice compositely includes all partial orientations as equivalent aspects of a whole mathematical entity.

The paradoxical facet also shows how the first section combines the different tensions into a three dimensional unity we have called *The Dice of Existence*.

The Hamilton rules for quaternions are:

$$i^2=j^2=k^2=-1 \tag{a}$$

In terms of our chapters, what we mean by this is that each paradoxical tension applied to itself realises something real. The paradoxical tenets of our first three chapters result in theories about the real world: quantum, relativity and electromagnetism.

$$ij=k \quad ji=-k \qquad jk=i \quad kj=-I \qquad ki=j \quad ik=-j \tag{b}$$

The second rule says how each combination of two elements infers the third. This quality of exchangeability of propositions gives the hub its rotational agility within the wheel of physics, where the dimensions naturally turn about a cycle of meaning.

$$ijk=-1 \tag{c}$$

The coherent ordering of the paradoxical dimensions into unity reflect how the Dice sections of this book are a three dimensional summation of the separate paradoxes we have explored in each chapter. The unity that emerges over the whole contemplation of separate paradoxes, does not remain with the individual ambiguities alone, but leaps compositely to a single real form. Paradox opens itself up, so that a whole meaning emerges as the outcome.

The book explores the quality of our experience in the world. But how it does this is to allow these individual chapters to turn on the dimensions of their paradox, so the rim of the wheel of physics as a whole may roll with the fulfilment of renewal over the separate sections. Physics is opened out to be the hub of paradoxical opposition about which the rim of meaning rolls.

There are two cycles in the wheel of physics. An outer circle of paradoxical logic fulfils in broad aim the specific happenings of the inner circle of logical consequence. The outer circle of meaning gives impulse towards completion, while the inner circle obeys causal inference. The two tensions of future completion and past objectivity meet in an event moment.

Event moment

Kauffman defines an event moment in Einstein's relativity. An event moment is not simply a point relating to past and future through light. An event moment is a structure of advanced half-wave from the future and retarded half-wave from the past, included in the describing of a place of meaning where content is added to make distinct the whole journey.

The conceptual description of the relation of time to space consists of a series of negatives of half-reality that develop on their journey into the meaning given by light. The event moment lives in a rhythm of its own making, as in the African tradition of the 'praise poem' drawing from context the route to an essential meaning. As McClure describes: 'The praise poems are ... created so that the young people in the tribe know who they are, who their ancestors are, why they are loved, and what special gifts they bring to the tribe and the world.' (McClure 2015) The conceptual description of the journey enfolds its verses into an encounter that is a single bestowal of meaning.

The potential of relationship of time to space converges inward upon the inevitability of light falling upon its own discovery of itself. The future delivers the meaning capable of stating the actors of the present in signified relation. The event moment is a going inward into the secret of the identity that reveals itself from the temporal context. Nothing has to happen at the event moment, except that the rhythm of reality describes its own dance into form.

Instead of defining the observer beforehand as a fixture of the reality to be explored, the observer is left open as a freedom to be resolved in the event, giving meaning to its relation to the observed. Similarly, instead of the observed being an objective truth already out there waiting to be passively seen, the material universe of objectivity is to be a freedom seeking expression in the meeting with the observer.

A form for holding these freedoms has already been laid down in the advanced and retarded half-waves adding up to a whole wave.

The advanced half-wave sketches the possibility of future time as a shaper of the rhyme, the balancer of lines. The retarded half-wave jumps in with the causal description of a story running through. These two impossibilities, from any classical perspective of history and prophecy, form between them the vessel for unprecedented meaning to find inevitable articulation. Out of all the stories ever told, of causality or imagination, the verse uniquely speaks a character of time.

The event moment gives identity of meaning into the developing of time backward, falling into the character, filling the empty story. The prophetic and the causal, so unwieldy on their own, lend themselves together to their natural meaningful unity. The two half-realities are blended to the single exceptional conjunction of their opposite versions of time into the miracle of the most peculiar rhyme. Their paradox is joined at the limit of what is possible, being spoken in time.

Causality and prophecy mark the beginning and end of every line. What happens within is revealed through the sequence of event moments that refine inwards towards the essential character that enlivens time. For through this descent into the inevitability of form, one meets that place where character speaks the spirit in the context of rhyme.

In all these examples of the freeing of time from an exact account, we understand the form through the journey. The elements of journey resolve their freedom in a form, establishing a relatedness of process, describing time.

Two half-dreams of the scientist and prophet now let fall between them a whole character drawn from time, to walk in time anew. As Lou Kauffman summarises:

I believe that what we see here is the beginning of an arcing connection between myth and science. Both find their commonality in the culture in which we are embedded and the language that we use. It should not be too astonishing that a descent from physics through its mathematical formalism to pattern and form leads inevitably to the mythology of the world arising – with the observer as imaginary mirror of the internal and external (which are really one). (Kauffman 1982, p. 124)

Spencer Brown reflects on the meaning that the process of engaging in his logic accomplishes:

The fact that, in a book, we have to use words and other signs in an attempt to express what the use of words and other signs has hitherto obscured, tends to make demands of an extraordinary nature on both writer and reader, and I am conscious, on my side, of how imperfectly I succeed in rising to them. But at least in the process of understanding the task, I have become aware that what I am saying has nothing to do with me, or anyone else, at the personal level. It, as it were, records itself, *that which is* so recorded is not a matter of opinion. The only credit I feel entitled to accept in respect of it is for the instrumental labour of making a record that may be articulate and coherent enough to be understood in its temporal context. (Spencer Brown 2009, p. *xx–xxi.*)

Spencer Brown, Kauffman:

◄———————————— emptiness/form ————————————►

Creation

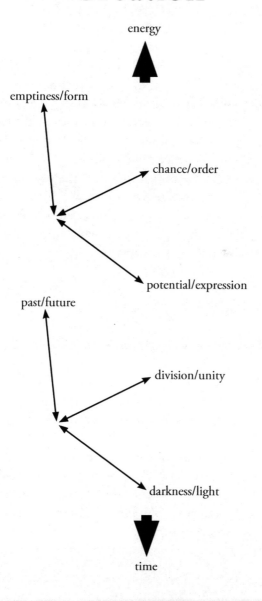

Chapter 7
Energy faces time: myth

In the meeting of Bohr and Heisenberg, we saw matter and meaning annihilate each other, unravel the fabric of meaning that could be made from matter, atom by atom. Von Weizsäcker, also involved in the Copenhagen meeting, transformed the destruction of the atomic split into a philosophical re-visioning of quantum theory. Unity, his experience taught him, had to be reintegrated as a basic tenet of the physics of existence. His research uncovered the need for returning the energy of existence into the context of happening. The exhausted past is re-energised in the provision of new meaning. Thus our chapter begins again with the experience of the journey to visit Bohr in Copenhagen.

Von Weizsäcker

Carl von Weizsäcker was an unnoticed traveller on the train with his colleague Heisenberg going to Copenhagen to meet Bohr in 1941 (described in Chapter 1). Von Weizsäcker visited the Institute in Copenhagen and had lunch there with colleagues, but was not invited by Bohr to his home for the discussions that took place with Heisenberg alone.

Even the two German colleagues probably did not know their motives in going to see Bohr. They were looking to talk to Bohr as a physicist to give definition to the project of constructing an atomic bomb. Standing at the centre of German physics, maybe they still felt the world could unify about some kind of centralised authority. Swept away in the certainty of their thought, they completely underestimated how Bohr, half-Jewish and shortly to flee Denmark, would react to their visit. The atom split not just theoretically but figuratively. The atom was no longer the shore of certainty towards which science had been heading but split apart in the very meeting to find resolution to the conflict.

When going to see Bohr, von Weizsäcker and Heisenberg must have felt themselves swept along in a movement that, despite everything, was going to secure the world on absolute fundamentals. They were caught in the momentum of their thought about what the world could be. They did not see the position of Bohr, at all.

Something in this journey is being spoken about surrender. One arrives at the knowing of meaning by giving up the medium of meaning in which one lives. Meaning is everywhere. We confuse the things we understand with the world as it expresses itself to us. Our science has knotted us in circles, telling us that it has the key to every meaning. We doubt everything unless knowledge can verify existence.

At the level of the universe, some question greater than the individuals involved is being asked. The theory does not roll up the universe as a carpet to be put away on a shelf. The theory asks the question of this essential greater task awaiting humankind.

Inconsistency

Von Weizsäcker wrote the book called *The Unity of Nature* in which he discusses the nature of time. Even reading his style, one is aware of an intellectual mind with great knowledge of philosophy, language and physics, pursuing meaning over many different frontiers. There is a brilliance of insight in his summation of the mathematical reality.

Von Weizsäcker, in his long career as a philosopher of science after the war, reflected on the way that the quantum theory he helped develop, carried an inconsistency.

> [In] quantum theory, we are made aware of an inconsistency in the usual presentation of that theory. The quantities that characterise

an object (the state vector in the Schrödinger, the operators in the Heisenberg picture) are written as functions of a parameter t, which is identified with time. Although, considered measurable in principle, time is the only measurable quantity to which no operator corresponds. Some other observable is always measured in the place of time. (von Weizsäcker, p. 397)

How can it be that in all the effort of understanding quantum theory, something so essential as the meaning of time, could have been overlooked?

Ungrounded time

In trying to reduce the world to static fundamentals, the one thing we forget is time. We assume that time is just a secondary by-product of the progress of things in the inevitable march of interaction to new states. Time simply allows happening to fall over its old states into new ones, in the tide of causal progression. But one forgets that physics has moved on, to the point in quantum theory where things do not have any prescribed, definite existence of their own.

All one knows of a system is the moment in which a question is asked. On this basis alone, a fundamental science is constructed.

The concept of a physical object (or 'system') presupposed by quantum theory is replete with problems. An object seems to be an object for a subject; the fact that the observer is himself part of the objective world is accepted as fundamental. (von Weizsäcker, p. 190)

A modern analysis of this state of affairs is given by Gambini in the advocacy of the rival *Montevideo Interpretation of Quantum Theory*.

The orthodox response of the Copenhagen interpretation argues that the objective of quantum mechanics is not to describe what is but what we observe. The measuring devices are classical objects through which we acquire knowledge of the quantum world. (Gambini *et al.*, p. 1)

So if the things themselves are not really there, until seen by an observer or measurement, then how can these same things be taken as the basis of time? But time is a fundamental parameter of the Schrödinger equation on which everything about quantum theory is calculated. So where does this time come from? And shouldn't time be treated as all the other variables, in some way derivative from the context in which it occurs?

One comes to a fundamental paradox. Properties refer to objects as static isolatable identities. But in as far as properties describe static isolatable objects, they do not relate to time, which requires a reference outside itself to qualify the behaviour of change.

Von Weizsäcker describes this:

> The isolation of an object one wishes to hold on to as a unity nullifies the measurement interactions required for the determination of its states in time. A strictly isolated object does not exist in time either. This nullifies the meaning of the fundamental concepts of quantum theory, and in particular that of the probability concept – of all the concepts, in other words, with which we formally describe an isolated object. (von Weizsäcker, p. 397)

Von Weizsäcker's loophole in the theory is the question of time and its relation to unity. If quantum theory deals only with isolatable objects and these are known only through measurement, then on what basis is time included in the fundamental equation in which these static entities are considered? The missing piece in the jigsaw of quantum theory is the relation to time. As Gambini continues:

> Time is treated unlike any other variable in quantum mechanics. The usual point of view is that to associate time with a quantum variable is impossible. (Gambini *et al.*, p. 2)

Jim Hartle, Professor of Physics in California, in an article *Excess Baggage* expresses this as follows:

Time is the only observable for which there are no interfering alternatives (as a measurement of momentum is an interfering alternative for a measurement of position). Time is the sole observable not represented in the theory as an operator but rather enters the Schrödinger equation as a parameter describing evolution. (Hartle, p. 6)

The analysis above admits that time cannot be taken as a principal parameter of mathematical neutrality, but must be derived from the relation of the finite behaviours to a unity that is beyond the elements themselves. That is, we must understand time as a quality of the theory itself or else the theory is nothing ('nullifies the meaning of the fundamental concepts of quantum theory').

The whole picture of quantum theory is founded on time. This picture falls apart because we haven't defined what time is and yet it is central to the whole picture we have built up. This led to the idea of von Weizsäcker and others that we need to deal with another notion of time. The current notion of time is simply something that goes from one moment to the next, the past is past and has no reference to the present, the future is still to come and the present will eventually get there. A new notion of time is needed, sufficient to illuminate what is happening in the atom.

Paradox of time

If everything understands itself as the viable transitions revealed to measurement, then on what basis is time included in the theory?

Lee Smolin writes:

Can a sensible dynamical theory be formulated that does not depend on an absolute background space or time? Can quantum mechanics be understood in a way that does not require the existence of a classical Observer outside the system? (Smolin, 1988)

Hartle suggests that:

> The notion of a preferred time in quantum mechanics is another case of excess baggage. (Hartle, p. 6)

Hartle argues for:

> ... a framework which dispensed with the excess baggage that the laws of physics were separate from our observations of the universe, a framework in which the inductive process of constructing laws about the universe was described in it, and in which our theories were seen as but one possibility among many. (Hartle, p. 8)

Something else has to be included in the expressions of the theory not directly measurable, from which the inner notion of time as a basis for change is generated.

Instead of seeing emptiness as a passive blank slate, we give an elementary property to the void. By giving emptiness this uneven character, even though there is still nothing in it, then only by participation is the construction of anything from emptiness to be understood. The emptiness is given place as an action that challenges existence to demonstrate the act of establishing itself. Subtly, in including emptiness, the goal of physics changes from explanation of what is static, to the proving of existence as a self-realising fulfilment. The emptiness actively clears the way, that the self-actualising inner core is freed to deliver its meaning in a world open for new statement.

Quantum time

We can understand quantum theory as following this primary movement from pre-unity (the generality of potential) to one as expression (the specific of existence, as described in Chapter 5).

Quantum reality describes the entirety of the domain of potential to expression, while classical physics concentrates on the level of already realised existence.

Quantum domain of description   **Classical description**

Pre-unity   one as expression   objective existence

(< - - - - - - - - - - - - - >)   (<>)

To say we want to treat time as a quantum variable means that we do not treat time as something external to the system. Rather we have to introduce time as part of the fabric of the system.

What we mean by a quantum process is something that is indefinite. What quantum theory is about, is considering these indefinite entities that are not really particles and are not really waves, they have no dimensions for themselves. There is no solid thing. In the Copenhagen interpretation, this indefiniteness exists in the beginning, but in the act of measurement this indefiniteness collapses down into something definite and fixed. Despite this indefiniteness, it is the nature of the world that what we see is something definite and fixed. So there is this indefiniteness, but luckily, at the act of measurement something definite and fixed appears. Quantum theory's hypothesis is that we can still see the world as made up of material atoms, definite things. We get to the foundation of material existence, even though that doesn't really exist.

Gambini and co-authors in the Montevideo interpretation suggest a participatory foundation to time as follows. They allow that the measurement of time itself has a quantum uncertainty, which gradually makes fuzzy the picture of happening, until the point where it is impossible to tell whether one is in a left hand side quantum description of potential to expression, or a right hand side objective description of the world. In other words, time becomes the variable that loses the indefinite aspect of potential by exposure to event. Potential erodes in the time of happening until only an objective reality remains. One no longer needs to consider measurement, for the blurriness of time is what pushes the system to state itself unequivocally in a given circumstance.

More generally, the effect of exposure to context enables a test to distinguish existence from emptiness in a recognised meaning, which cannot be differentiated from an established identity.

We can illustrate this with an earlier example (described in Chapter 5). In the case of the bacteria, their basic movement from potential to expression occurs in two radically different ways. In one way bacteria act in the world as individual entities, in the other they cooperate as a colony, united physically and functionally. What the Montevideo interpretation is saying is that time begins to happen when individual bacteria act out their response to circumstance in an ambiguous way. Say, that in feeling hungry, they send out a chemical signal (cAMP) to other bacteria. Then the various responses that take up the ambiguity of this message in the other bacteria, gradually turn this signal of potential, into a situation where one or other meaningful behaviour is expressed over the whole colony. The time of happening gradually absorbs the potential into one or other action of behaviour.

The bacteria have different definite states they can occupy. How can these single celled existences form into a colony with billions of bacteria acting as one organism? Are they super intelligent with communication and organisation abilities far surpassing the human capability? It makes more sense to imagine the bacteria in a liminal undecided state, with these potentials inside them, to exist as individuals or as a colony. According to context, the indefinite state becomes something real. Bacteria have a quality of existing in a state betwixt and between.

Time, according to the Montevideo interpretation, as experienced by the bacteria, would be this concentration of all possibility, with many alternate meanings open, into an imperative of action that impressed its urgency with greater and greater insistence. Time would extract from the potential of everything that could happen, the outcome of one or other behaviour, as fulfilled meaning. Time would do this by, gradually in its description of event, translating the potential into a single choice of expression of rounded meaning.

In the case of the slime mould, also considered in Chapter 5, the collective state acts exactly as a single existence made of cells. This composite organ moves off to some suitable place. It then makes a fruiting body. This enables some of the member amoebae to sporulate from the branches of its own collective structure. The onset of sustainable conditions returns the amoebae to single-element entities with potential to cooperate once more.

Quantum time induces the choice of a collective state so that the amoebae manifest purely as one organism. On fulfilling the cycle, the free choice of potential is allowed in the life of the amoebae once more.

The potential of many individual choices supports a meaning in which the potential of time takes on a specific character by focusing indeterminacy upon a definite form. This time fulfils the requirement of von Weizsäcker and Gambini of being a quantum time, whose meaning arises in a practical fulfilment of an individual potential. Time states itself unequivocally only when the ambiguity of individual elements settles upon a meaning.

However the theory leaves time as a concept still imprisoned within the treatment of energy! Happening takes energy from indefinite states to actual wholes, but time has no say in the matter! Time merely achieves an energetic transformation from indefinite state to definite existence. The play of time remains a shuffling of the energy within the finite realm.

Relation to the infinite

Von Weizsäcker gives us deep philosophical roots into the very progress of science for how the oversight of time came about. The key to the loss of quantum time he identifies, is in the displacing of the infinite from the potential of things. By limiting what we understand as existence only to what is expressed, time is perceived as something external to existence.

To conceive of the infinite as potential … avoids the paradoxes that critical minds have discovered again and again in the conception of an actually existing infinite.

Anyone who has understood this may wonder why mathematics nevertheless shifted in the second half of the nineteenth century to the conception of the infinite as actual, which is so much more difficult to justify. So successful has this shift been that is nearly impossible to disabuse the contemporary student of mathematics of the superstition that this conception is the only one possible. (von Weizsäcker, p. 346–347)

For Aristotle in Greek time, and Kant and Gauss in modern philosophy and physics, the infinite was understood as potential. There was then a natural reference to time beyond its own finite measure. For potential to become actual, time was found in the relation of the finite to the infinite. Time was not a parameter of measure of a controlled process, but a mediation of the finite to the potential of the infinite which always hovered as influence above.

Nineteenth century mathematics imputed that actually everything, including infinity was graspable within a single comprehensible theory (as described in Chapter 6).

The relation to the infinite is crucial in fulfilling this project of establishing a true quantum time that lives in the potential of things to become.

Archetype

After a breakdown in 1931, Pauli, in communication with Jung, started to dream in terms of physics about the healing of the world. He dreamt that what the atom represented wasn't about something impenetrable to do with measurement, but it was about this other nature of time and experience. How does that experience of us being in the world, between wholeness and separation, reveal itself? The atom was in between time and energy, in between potential and existence. Instead of looking at it as the basis of matter, his dreams were showing him you could use the same foundation of physics to understand how time could reveal itself in archetypal form. It turned the atom around so that instead of it saying how matter existed independent of time, it showed a model of how existence could reveal itself in a unique statement of its becoming.

The atom wasn't just telling us about static matter. It was also telling us about dynamic meaning. It was telling us something about how a quality of experience is able to distil itself into this archetypal moment of revelation. The dreams were using exactly the same symbolism as Pauli was working with in his day life, but the transformations, qualities and nucleus were all about how meaning distilled itself. There were two dimensions: either you could interpret the atom as a stability that rode the waves of time and always remained itself; or, there were these transformations in happening. The transformations had this energetic quality of arriving at something concentrated and distilled which he identified with the archetypes. There was an ambiguity at the very centre of the world, that matter and meaning were not separate. We could read the atom in two different ways: one that jumped across the waves of time and had this stable form, and the other which had the quality of allowing the transformation of experience so the atom could describe how that transparency of experience could manifest itself as distilled meaning of the archetypal form of arising.

Bohr used the ambiguity between stability and happening to say the atom is the form of something that is able to ride the waves of time and remain itself. But Pauli said we could understand this differently. For this model of how things happen to us, how things unfolded in his dreams, is very like the way experience as a fluid thing is able to crystallise itself into meaning, a quality that reveals what that experience signifies. His alternative view was that this model also carries meaning.

Indefinite hours

In October 1938, Pauli sent Jung a dream on time, very similar to the observation of von Weizsäcker. He introduces the dream by saying:

> I have come to accept the existence of deeper spiritual layers that cannot be adequately defined by the conventional concept of time. The logical consequence of this is: that the death of the single individual in these layers does not have its usual meaning, for they always go beyond personal life. In the absence of appropriate terms of reference, these spiritual spheres are represented by *symbols;* in my case, they are particularly often represented by wave or oscillation symbols (which still remains to be explained). (Meier, p. 21)

In the dream of January 23, 1938, two waves of temporal cycles are shown one under the other to give a window on the universe.

> At the top there is a window, to the right of it is a clock. In the dream I draw an oscillation process beneath the window – actually two oscillations, one beneath the other. By turning to the right from the curves, I try to see the time on the clock. But the dial is too high so that doesn't work.

Then the dream continues. The 'dark unknown woman' appears. She is crying because she wants to write a book but cannot find a publisher for it. In this book there is apparently a great deal of material on time symbolism – e.g. how a period of time is constituted when certain symbols appear in it. And at the end of one page of the book there are the following words, read aloud by the 'voice'.

> The definite hours have to be paid for with the definite life, the indefinite hours have to be paid for with the indefinite life. (Meier, p. 175)

In January 1939 Pauli further elaborated:

> I think I can take the matter a little further. The dream affords a certain insight into how a periodic symbol actually comes about, for in it a dark man appears (associated by me with the 'shadow' archetype), who is cutting pieces out of a zone of light at regular intervals. My own idea about this (arrived at purely intuitively) is that these symbols have a connection with the *attitude to death*, in that one oscillation signifies one human life, but one which is to be interpreted only as part of a larger whole. (Meier, p. 24)

The dying of the individual within a greater significance of life is a doorway into this other notion of time.

Pauli's interpretation of the cycle of death and life is of this surrender of the wholeness of identity, into a plenitude that life precisely fulfils. The individual dies ('one oscillation signifies one human life') but only to realise the surrendered plenitude of wholeness ('to be interpreted only as part of a larger whole.') The finite fulfils itself in dying ('these symbols have a connection with the attitude to death') by becoming the relation to the wholeness that thereby achieves a periodic fulfilment.

Ancient root

Von Weizsäcker towards the end of *The Unity of Nature* makes a throwaway comment:

> Being can only be in time. One will have to distinguish the time which abides in the One (*aion*) from its copy (*chronos*) which progresses in accordance with number and is counted by the celestial movements. (von Weizsäcker, p. 398)

The root of chronology is the Greek word *chronos*, which refers to the regular rhythms marked, say by the planetary cycles. *Chronos* is very familiar to us in modern hurried existence, as the outer linear timescale of deadlines, plans and appointments that always seems to be filled with busy-ness no matter how hard we try. The Greeks sensibly, as true to their nature, had another word *aion* meaning unbounded time, which is also translated as different expressions or eras of an eternal time.

Chronos is the name for the chronological rhythms, of the earth moving around the sun, as an abstracted external motion: very useful to plan a meeting or organising when to sow the crops, marking the passage of time, in a routine way. But there also is another word for time, *aion*. And *aion* is unbounded time, or mythological time. *Aion* time isn't something we can reduce to an external measure. It is something we have to participate in.

Aion time is about relaxing into a happening that is synonymous with meaning. The challenge of von Weizsäcker is to move from *chronos* which is always full of physical things happening into *aion*, which is open to the domain of meaning. This is a big ask for physics!

In *aion* time, the dynamic and the unity are both there and present, together. What signifies the dynamic process is the *meaning* that enters into the vessel of the isolated parts, bringing them together. The *meaning* involves the various structural parts in a movement that relates them all.

The happening cannot be separated from its meaning.

The relation of *aion* to *chronos* time is seen from the hub and the rim analogy of Chapter 6. *Aion* time is the rim that turns about the universal meaning, while *chronos* time is the hub that keeps order in the basic structure of existence.

Holistic approach

The challenge is to change the way we use time in relation to the atom, from simply an account of material happening, into the disclosure of significance.

In *chronos* time, the atomic description is of a static thing, the basic element of matter that is placed within a medium of happening, denoted abstractly by the variable 't'. The use of *chronos* for time is atypical to the approach of quantum theory. One would expect the variable to take breath in the defining of (and being defined by) other variables, more as *aion* time.

To gently explore this, we return to Hamilton who we already met as the discoverer of quaternions in Chapter 6. Hamilton with his colleague Jacobi also created a holistic method for studying the dynamic progression of an energetic system.

The first thing Hamilton-Jacobi do when faced with a physics problem, is to take a breath and consider the whole. Instead of focusing on the individual particles, the Hamilton-Jacobi description looks at the energy of the system as a whole and how this varies with time. From this overview, the approach then concentrates on the individual articulations of matter. The various different wave possibilities of how potential actualises itself in space and time interfere with each other, to give a localised path (see Misner *et al.*, p. 641–43). This localised energy/event is identified with the particle.

Instead of beginning with an indefinite particle and then trying to work out how interactions are going to develop (as we explored above in quantum time), the starting point is to oversee the total system energy varying in time and from there to arrive at the locality of the particle.

The Hamilton-Jacobi description allows us to accommodate *aion* time into the interpretation of quantum theory. What we are interested in discovering in *aion* time is the meaning. We do not need a focused interpretation of either the particular expression of energy or the specific detail of time. What we do want to focus on is how energy and time as possibilities, mediate together a full articulation of meaning, i.e. the convergence of content and happening on a form.

The content and the medium of happening in this approach are not defined at the beginning, but take on a form in the way the generic potential of energy and time realises itself into actualised existence. The *aion* perspective brackets together {energy, time} as two related mediums in paradoxical tension. The *aion* perspective asks, 'What is the energetic requirement and context of event, in which meaning may distinguish itself?' The realised meaning actualises energy and time, as the medium for the distinguishing to occur.

Earlier in the Montevideo approach, time was constrained to resolve the indefiniteness of energy. In the *aion* approach, the starting point is that *both* energy and time are indefinite. The exposure to context draws out of this ambiguity, the specification of a rounded energy appearing through a particular path of happening. Energy and time are defined in the way their paradoxical entanglement finds, in context, the means of differentiation of a separate existence.

First light

Time in its understanding within physics is purely believed to be about existence. Time deposited us on this shore from which we face forward to the future. The necessary assumption for this view of time is that we began in a highly ordered place. Existence was super organised, life-like, with an inordinate amount of implicit structure. In the gradual dying of this concentrated state, existence happened by material chance upon the disordering into the parts that we now call galaxies and planets and life and things.

This argument is itself however disputed. Roger Penrose in his book *Cycles of Time* makes the point that a Big Bang is far from a neat and tidy event. To suggest that such an origin had this immense order of a singularity that could materially wear down into a universe of specialisation is highly conjectural. Penrose begins his argument with an imagined dialogue of a child and an adult.

Where does this thing you call 'organisation', whatever it is, *originally* come from?

–Ultimately it comes from the Big Bang, which was what started the whole universe off with an utterly stupendous explosion.

A thing like a big walloping explosion doesn't sound like something organised.

−It's one of the biggest puzzles of cosmology where the organisation comes from, and in what way the Big Bang really represents organisation in any case. (Penrose, p. 5)

In taking a detailed picture of the background radiation, the recently completed WMAP (*Wilkinson Microwave Anisotropy Probe*) observation made out tiny fluctuations in the way the universe had developed, which could be mapped on to the location of galaxies as we find them now. But this also raised a fundamental question.

Quantum theory only moves from its state of collective potential when the system is observed. For the potentiality of the universe to turn into the actuality of an actual distribution, something must be doing the observing.

But in the early quantum universe these conditions were not present. So it is unclear in this model how materialism is responsible for seeding in its natural flow, the wholeness that we see reflected from an early state of the universe. It is in a way surprising that with our modern technology we get such clear pictures of something so fundamental as a Big Bang developing in some kind of recognisable progression to the universe we see now. As Chris Clarke writes:

The WMAP satellite observations of the universe at an age of some 380,000 years confirm a picture in which the universe has evolved as if it started in a perfectly smooth homogenous state. By the epoch observed by WMAP we see minute fluctuations superimposed on this uniform background, of the same character as the quantum fluctuations that can be detected when a uniform beam of radiation is observed in the laboratory. On conventional theory, these cosmological fluctuations grew under the influence of gravity to produce stars, galaxies and ourselves. Note, however, that in quantum theory it is the act of observation that precipitates quantum fluctuations: without observation (in whatever generalised form we may conceive it) a homogeneous initial state evolving under homogeneous laws must remain homogeneous. So the early fluctuations that eventually give rise to the existence of planets, people and WMAP are caused by observations such as those made by people and WMAP! The problem of quantum observation lies at the heart of modern cosmology. (Clarke 2011)

The entanglement of {energy/time}, which characterises *aion* time, is already present in any mathematics trying to argue something from nothing. For instance we already came across Eddington's refusal to accept his own solution to the black/white hole singularity (essentially what a Big Bang event is) because his solution entangles matter and time, in a way he felt invalidated physics as an explanation of what was there. But when we begin with an entangled {energy/time} as our basic model, then nothing and everything, both remain continually in reach of the definition of existence as a temporally defined universe.

Classically, as in Hawking's book *A Brief History of Time*, we imagine energy as primary and time as the mere record of what happened to it. But when we elevate {energy/time} to an entangled possibility, then happening exists in an exploratory state that is understood as a sequence only in coming upon the energy of a defining meaning!

In fact this is our experience of creation, in the small ways we encounter it, where the moment of meaning seems an energetic event that orders everything we have been doing up to then into a fulfilled sequence. We are struggling to achieve something that does not seem to want to happen. Then at a particular moment, a subtle change comes about. The happening opens to rest in the realisation of a meaning. The whole meaning resolves a latent potential of energy, to express itself in a fulfilled time.

In the same way, the energy of the galaxies was present as a potential that realised the actual matter of stars and planets, only in resolving the playful happening into a fulfilled origination.

The entangled state of {energy/time} realises itself as a meaning in which energy as distinguished matter is stated within a history of separated time. Such a meaning of matter of the universe, governed by a history of arising, seems to hold us in a unique solution to existence. However the very nature of the entanglement of {energy/time} implies that such a formulation can have other resolutions of meaning! Moreover we are involved in giving this meaning.

Creation

In 1953 Fred Hoyle, a British astronomer, was considering the production of the element carbon, essential for life. The chemical elements after hydrogen and helium that would have initially formed at the Big Bang were made in the hot centre of stars and then ejected across the whole universe, so arriving at earth.

The production of carbon from helium and beryllium competed with the reaction of the carbon and helium to make oxygen. Hoyle understood that the carbon had to somehow promote itself within this pathway of reactions, so that enough carbon could be produced as to enable earth to have this element as a fundamental resource for life.

Being an astronomer and without knowledge of any of the complicated nuclear physics involved, he predicted that the only way carbon could create itself in the quantities necessary for life, was if its nucleus had a particular excited energy state, resonant with the combined energy state of helium and beryllium. Now there was no reason at all why it should have this exactly fitting energy state relative to the energy states of the two producing elements; the sole argument Hoyle had was that this was the only possible value that would allow life. Testing his prediction, physicists duly found the excited energy state that Hoyle predicted, with the ridiculous, precise number of 7.6549 MeV.

For the process to work and for human beings to exist, carbon had to possess an excited state with an energy almost exactly equal to the combined energy of a beryllium nucleus and a helium nucleus. The combined energy of beryllium and helium is actually 7.3667 MeV – just below the 7.6549 MeV energy state in carbon. However, the temperature in a red giant star is high enough to boost the energy of motion of the two nuclei to just above the critical threshold. For carbon to be created, the excited energy state of carbon had to be just right, and so did the temperature inside the red giant. It was an extraordinary coincidence that they were. (Chown, p. 177)

Further it was necessary that oxygen should *not* have an energy state that would have decomposed the carbon as it formed.

Had the energy state of oxygen been a fraction above 7.1616 MeV instead of a fraction below, there would hardly be any carbon in the universe.

Instead, life is possible because nature has fine-tuned the properties of three different atomic nuclei. Beryllium is unusually long-lived for an unstable nucleus. Carbon possesses the exact energy state needed to promote its production. And Oxygen lacks an energy state that promotes its production at the expense of carbon. (Chown, p. 178)

Following from this prediction of Hoyle of properties of nuclei understood by the necessary relations that allowed the requisite elements to be formed inside stars, the question was raised: how does the universe possess such fine-tuned parameters that life could have evolved within it to observe it?

The remarkable series of nuclear coincidences first recognised by Hoyle was to have a profound influence on his life. It convinced him that the universe was geared up for the emergence of living organisms. Life had arisen on earth not because of some zillion-to-one accident but because it was truly a cosmic phenomenon. (Chown, p. 178)

To make sense of the *fitting* together of the energy states of the three elements, Hoyle moved towards acknowledging an interconnected meaning in everything. One had to place the relationships of the three energy states, perfect for realising life, within the context of the overall meaning that this synchronisation of values brought about. The meaning (of which Hoyle's consciousness was one aspect) was exactly what unified the separate possibilities of the early universe, into a march of time, i.e. the collection of events happening into a universal reference of unfolding. The meaning determined that time was more than just isolated, spontaneous and random occurrence.

The entangled state of {energy/time} resolved into the meaning of galaxies, and carbon and life, as distinct material manifestations, within a time description that joined all these events together, from the early creation of elements in stars, to the production of life, contemplating the universe. Such an expression of meaning is not unique. We may also respond to the {energy/time} entanglement by nullifying the potential and erasing time.

Destruction

Von Weizsäcker worked on the calculation of the weights of the atom including a formula for what is known as the binding energy. When the weights of the atomic elements were explored by Francis Aston in the 1930s using a mass spectrograph, it was found each atomic element had a unique binding energy.

The binding energy is a favouring of the whole elemental state beyond the existence of the individual particles of the nucleus. Each element possesses its own binding energy (or incentive to stick together as a whole nucleus, rather than to fragment into parts). This eating away of a small amount of the total energy of the atom from the particles independently is different for each atomic element.

Each elemental frame holding together the particles is slightly different. In transformative processes of one element into another, since each element possesses its own unique binding energy, this process need not be energy neutral. It can be that some mass is left over from the exchange.

For instance, in the sun, fusion of helium is the source of the intense temperatures that warm the earth. Heavier elements in undergoing nuclear fission into smaller elements release an excess of energy. What stimulated von Weizsäcker and Heisenberg's trip to see Bohr was the realisation that this process could be artificially stimulated. By firing high energy products from radioactive decay at the nucleus of an element such as uranium, one could trigger a cascade into smaller elements artificially, with a remainder of energy being released.

It was realised that nature favoured the process of transmutation of the elements. In the changing of the atomic structure, two neutron particles (used initially to fire at the nucleus to disturb its configuration) were released as excess to the requirements of the new elements. These two neutrons were on hand to wreak further havoc down the line, entering two more hallowed spaces of the nucleus of other atoms, until as a pack of cards, the whole stock of available nuclei all changed dramatically in their own augmenting impulse.

In the atom when it splits, the anonymous symmetry plays back on itself. A neutron hits an atom. The nucleus breaks apart into smaller elements releasing energy and also two more neutrons. The picture is unchanged except that now two neutrons are around, looking for further destruction. The stage, impervious to the nature of the particles that inform the statistical law, is simply reset to allow the two neutrons the same play, two atoms to split and four neutrons to be freed.

When the first atomic test was done in the desert around Los Alamos, Oppenheimer the head of the programme, quoted from the *Bhagavad Gita*, 'If the radiance of a thousand suns were to burst at once into the sky, that would be like the splendour of the mighty one.'

Later in 1965 Oppenheimer recalled:

We knew the world would not be the same. A few people laughed, a few people cried. Most people were silent. I remembered the line from the Hindu scripture, the *Bhagavad Gita*; Vishnu is trying to persuade the Prince that he should do his duty, and to impress him, takes on his multi-armed form and says, 'Now I am become Death, the destroyer of worlds.' I suppose we all thought that, one way or another. (Oppenheimer 1965)

Myth

Aion time is about the creative impulse itself, how form comes into being. As with Goethe, the dynamic of the chestnut tree growing isn't something we can reduce or abstract or put into a physical theory. The dynamic of the tree growing is something we have to encounter. It is a journey the tree goes through every season. The tree is a journey, shaping the form of chestnut through all the gestures the chestnut takes on, to express some part of that journey. So in this other type of time, to understand the nature of the tree, we have to meet the dynamic quality that is the chestnut. The cycle from all the individual cells to the differentiated meaning of the chestnut tree is the quality of its time.

This participatory time has a completely different quality to it and it is also connected to the myths. The myths of creation aren't trying to understand something as an explanation, they are trying to tell a story. A story about how the universe came into being. That story has the quality of a bizarre narration, that in the beginning was darkness and light, void and form, chaos and meaning, and out of the mythological characters and their interplay, earth had come into being, life had come into being and man had come into being in tumultuous relation to the gods. In mythological time you aren't meant to understand it, set your watch by it, or grow the crops by it. This is the very story of unfolding and creation that we are in. That is the way that the universe in all its separate material elemental building blocks, has arisen into this form. This very process of giving birth is reflected upon in these myths and this other notion of time. It is about the creation of form and structure.

This brings us from the experience of *aion* time, as told in physics, to another apparently quite different realm, its role in the theology of the Old and New Testament! The original language of the Old Testament was Hebrew. When translated into Greek, the Hebrew word *olam* (a specific word for time), was translated as *aion*. The original language of the New Testament was Greek and *aion* is again liberally sprinkled through the text. Together the two words *olam* and *aion* and their adjectives appear more than 700 times. But the challenge comes when translating these words into English. The two words are often mistranslated as Eternity, there being no suitable other word for *aion* time. Eternity completes the job of the separation of energy and time in physics. It implies that the spiritual realm is beyond mortal time, in a quite different realm to anything that may go on here on earth!

Olam and *aion* signify that out of the entanglement of the potential of energy and happening, God appears together with the detailed description of how the dynamic, whole challenge is set in the midst of the events of daily life. This is not to place an isolated God within a *chronos* time. Rather God appears out of the veil of hidden meaning as that which acts in the line of a certain articulation of events into a story that is not yet fully told.

From the modern scientific perspective, the act of seeing is an essential element of the material-temporal universe. The observer in relativity or the measurer in quantum theory plays the freedom of chance in the mathematical formulation into the physics of a reliable structured reality. In a theistic understanding, the engagement that is needed to retrieve meaning implies always a perspective outside energy and time. The journey of darkness to illumination knows ultimately a witness, surrendering itself to the fate of a meaning, over which its existence presides.

Story of time

The Hindu God Vishnu, the Preserver, to whom Oppenheimer refers, is standing next to a scientist and says to him, 'I am curious, go and explore the nature of time.' The scientist goes off into the lab to do Vishnu's bidding. As one event leads to another, he gets more and more curious and soon is involved in the mystery of the double-slit experiments. He forgets the original question, 'What is the nature of time?' Instead he uses time as one of the variables in his theory. He starts gaining more and more knowledge about what happens to matter and energy as it develops under the conditions of the experiments. He completely forgets about time as the question he was asked. And he starts realising that electrons interfere, and there is entanglement and he gets more and more puffed up with his discovery. And completely forgotten in these theories is that there is the variable 't' for time. He is using time but has forgotten about the initial question.

Having developed this theory about the universe, the scientist works out that matter has this curious ability to split apart. You take one atom of plutonium and fire something at its nucleus and the plutonium splits apart and you get two smaller elements and some energy being given off. There can be a cascade of the splitting of these elements of matter and this gives off a huge amount of energy, each time this happens. The heavy element plutonium has more mass than the elements it breaks into, so the rest of the mass is given off as energy. Man the scientist, puffed up with his knowledge, then discovers you can make a bomb by this principle of splitting apart matter. And that bomb is capable of huge devastation because you can release a huge amount of energy from it. Man has gone down this road of destruction.

To try to address the possibility of making a bomb, Heisenberg, German, goes to visit Bohr, in occupied Denmark in 1941 (described in Chapter 1). Von Weizsäcker, after that momentous meeting in Copenhagen, later comes to reconsider the question Vishnu had asked the scientist right at the beginning. Even though everything is being considered in the equations, time is included as a variable with nobody thinking about it. And so von Weizsäcker says that what we need is a re-assimilation of time, a re-understanding of time. We shall have to decide whether we are going to stick with *chronos*, as a linear time, or go to the Greek idea of *aion*, which is a creative time, the time of mythology, the time of story. To make quantum theory consistent, we need to re-embrace this *aion* time. Because nothing has precision in quantum theory, everything is entangled and is a possibility. Yet time is the one thing that still has a precision, which is completely contrary to the theory. Von Weizsäcker says that what we need is a participatory time, the time of creation, the time of how new form comes into being. That time shouldn't be something that is external to us. It should be our own journey. It should be something that creates itself in the story. It should be the understanding of how the finite and the eternal, the individual and the universal, integrate with each other and produce a story of existence, that isn't separate to what exists. Just like the chestnut tree is inseparable to the story of how it produces the bud, produces the leaf, produces the flower and produces the seed, so we should be living again in a mythology of realising our dynamic relation to the universality of form. We are doing something together. We are not simply these isolated individuals doing our own thing. We are moving to something collectively, that is universal and embodied. This was a physicist writing about his understanding of physics, talking about the need to work with a different sense of time, which is the story of existence itself, coming into being, and becoming a movement addressing, how as individuals we come together as form.

We bring that quality of arising of a form into an ongoing story about who we are. The story doesn't try to reduce us to individuals and atoms and genes, but takes us further into who we are as a whole, expressed together. This quality of time cannot be reduced. It cannot be understood by abstraction or by physics. It is something that has to be encountered, like a religious story, like a traditional creation myth. The scientist gets lost in the knowledge that he can gain about a finite world. Then he meets a calamity and in that calamity he is brought back to the question he was asked. And the question he was asked was, 'What is that creative nature of time, the mythological story?' In having tried to understand the world through abstraction and finiteness, the journey into science brings us back into the question, 'What is the myth of our time, what is the story of creation that we are in?' Everything needs that story of creation. Everything in nature has that story of creation. It is not something we can reduce. It is something we have to participate in, giving us our direction, our collective aspiration to a universal form, something that holds us all. Our age is about Vishnu asking, 'What is the nature of time?'

We have forgotten where time in the equations came from. We have forgotten the treasure of time, which is what oriented us, hidden underneath the ground of all this mathematics and physics.

't' is still there but everyone has forgotten what it represented. We have taken away from ourselves the very thing we need to live, the very story that organises us and orders us.

Illuminating time

Evidence of this cycle of meaning working out of emptiness is found in the role played by dark matter and dark energy. Dark matter and dark energy were originally proposed as a theoretical construction to help explain how the universe does not fly off into dissipation or bunch up into over heavy clusters. Dark matter and energy were mere theoretical safeguards to artificially ensure that the matter of the universe could behave with the right mixture of freedom and cohesion. It was simply envisioned as a theoretical stopgap.

But now in our literal inquisitiveness, we are asking ever more detailed questions about 'what is it?' without getting any obvious replies.

The best answer seems to be that dark matter and dark energy are the necessary inputs that allow the universe to form within itself, cycles of renewing meaning. The presence of dark matter and dark energy enables the character of the universe to fulfil its own origin, without dissipation or collapse.

For instance, dark energy is traced to an arbitrary, almost vanishingly small, cosmological constant in Einstein's equations, whose only motivation seems to be that it holds the universe at just the right expansion rate. Dark matter in the bi-metric theory of relativity is argued as a future originating compensation that regulates the past progressing paths into a round of regularity.

Without the dark matter and dark energy, there would be no cycle to hold the meaning of the universe in rounds of periodic completion. Dark matter and energy are precisely what it takes to hold the universe in a cycle of transformation that fulfils the character of its own inception.

The Large Hadron Collider, amongst other facilities, is due in the current round of experiments starting in 2015 to make a pretty exhaustive search amongst matter itself for the source of dark matter and dark energy. My prediction is that no direct causal origin of explanation shall be forthcoming. Instead we need to embrace another type of matter, as *aion* is another type of time, in which unity advances its whole meaning through the course of the universe.

The question dark matter and energy are asking us is about meaning. We need to stop focusing on energy and time as precisely understood objects and events. Instead we have to address the generic quality of energy and time that safeguard the whole meaning of our conceptual reality.

Dark matter and dark energy challenge us as to whether the unity of existence is indeed buried away in the recesses of fragments of particles or whether the unity is something tangible that we all access, and to which our choices all relate. Dark matter and energy allow for a universe of existence to frame itself dynamically out of emptiness. More than this there is nothing.

In physics, when we pay attention to all the paradoxes in the journey of the universe, we arrive at the complementary notion of the meaning one would normally associate with theology and myth! In journeying through the paradoxes of physics, we are left with the suggestion of the unity that gives to finite events the cohesiveness of their universal telling.

Each culture that has founded itself on an address of meaning has developed a corresponding science, giving cosmic description to the events of creation. Heaven's meaning complements Earth's tale of elemental interaction in the description of science.

Nucleus

Our inquiry throughout this book has found a deeper motivation to science than only the material. Already from the first chapter we saw how Bohr had intuited this and employed the stopgap of identifying measurement as the motivation of matter.

One thing however that stood out as an anomaly was the nature of time. Time is not just about a passive backdrop for matter to take its course. In Greece, the interpretation of *chronos* time stood alongside *aion* time. In the latter, attention is shifted from matter to the motivation-to-be.

Throughout the writing of this book, my aim has been to bring back the focus of physics to the motivation-to-be. In each chapter we saw, how through division and unity's perspectives on the double-slit experiment, through Newton's and Leibniz's opposing views of physics, through retarded and advanced waves, there is always this pattern of polarities in our motivation-to-be that determines the physical world we describe. This motivation-to-be has, in the history of science, often been decided by a short term utilitarianism that chooses the deployment of time into energetic activity, over the long term contemplation of the exercise of whole fulfilment. At the culmination of this short-term understanding, science lives out the consequences of this motivation in the atomic bomb in which energy separated itself as the final anonymous product into which all time was reduced. The atom split into an ultimate destruction. But from where is this destruction visible? In relation to what other ground does the atom split?

In all my writing I have shown, through my experience, that motivation-to-be has this other long-term choice for whole potential, described in my travelling from Oxford to Africa, and from Rotterdam to the Alps. My motivation stimulated another form of experience. These forays into a different illumination of the world finally impelled me to make a deeper commitment to the Alps and this other way of experience. It was in the Alps I met with an energetic fulfilment of this journey.

In energy and time facing together, does the model of the atom we have used to hold energy also apply to the outcome of exploring our motivation-to-be? When free individuals form into a colony, as considered in Chapter 5, energy shapes the processes of time so that the impulse of behaviours cohere into a collective form. The nucleus settles the periphery archetypally of time's questions.

As described in Chapter 3, light is not really a thing, more a chance that happens in freedom between two mutually initiating modes of description: electric and magnetic. The nucleus anchors the forays of the outer electrons, into a huge stable energetic resource. The way we understand the nucleus is that its energetic capacity is held together by two new forces: the weak force and the strong force.

The weak force at the gateway of the nucleus has to it a double dimension of freedom whereas light has only one. In modern quantum theory this aspect of the foundation of energy is accepted without giving any physical motivation for this character of the weak force. However, once we place the energetic model within the motivation-to-be, then we recognise this double freedom as the polarities of interpretation that are always available to motivate existence in the world. This double freedom to motivate existence in the world is the basis of each chapter – between chance and order, potential and expression, emptiness and form. In surrender to the existence of these polarities, we are freed to go further, taking another step inward through paradox into a new order.

In that same primary motivation for wholeness, we find within the weak force, the strong force, where a third degree of freedom is present. The strong force is the founding motivation of any physical description, the all-powerful glue holding the central particles, the pervasive arguing of identity over freedom. Just as the strong force is breached to expel the energy of destruction in atomic physics, so too the strong force holds the subtle substantiation of our motivation-to-be in an anchoring energy towards creation.

The motivation of existence is anchored in the energising of form, to give collective reality to our individual searching. The double-dimension of holding the ambiguity of motivations in our journey around the rim, fulfils itself in the single statement at the structured centre of causal consequence, concentrated at the hub.

The splitting of the nucleus of the atom has been a shadow of destruction darkening our whole social outlook, and forcing us to ever placate the gods of science which we unwittingly awoke in such terrible fashion. It has led to a culture fixated on things, concepts and labels, as if by understanding everything, we might disarm the terrible threat at the centre of our knowledge. In this book we have travelled again over the ground of science, and found another way we can put together the story of the atom.

Our effort has been to shift the foundation of physics from the fixity of conceptual interpretation to the dynamic distinguishing of active meaning. In this latter approach, the nucleus is no longer something fixed that we have to control, and whose threat rules over us in fear. The nucleus is about holding the dynamic freedom of the world through the names by which we address wholeness.

In the mountains at the end of this story, my experience was asked to choose between two motivations of existence. Thought had taken me to a place of despair, in which destruction was the only logical consequence left as its outcome. But the world offered a second path, which was to engage anew with the landscape of feelings, people, adventure, as the formative ground of experience. I understood this choice as needing to descend from the mountaintop of specialisation, abstraction, isolation, and to enter the valleys of people, emotion and life!

In this confrontation of darkness and light, the world of possibility formed about its own identity of becoming. Something entered into a situation that was stuck, as if from outside, and resolved in the light of liberation of that whole 'aha!' sense, all the open questions in a new illumination. This moment of unlocking the puzzle was not simply thought opening up a box with a new conceptual map. It was experience finding in the world the existential ground of being, as the basis on which all happening is predicated.

We do not have to be frightened anymore by technology as if our mind has unleashed some artificial monster of destruction we have to control. For the basis of our science, as Goethe showed, is about the moment of resolution in which experience is illuminated in the worth of all that is. It is to experience that we need to look to know the pattern of the world!

Time happens upon a ground that is given energetic pattern by an active process of formation. *Light* illuminates the whole being, at the centre of all happening. Energy orders the *dice* of experience, in the naming of *creation*.

Bibliography

Aston, Francis (1933) *Mass Spectra and Isotopes,* Edward Arnold & Co., London.

Barfield, O. (2011) *Saving the Appearances, A Study in Idolatry,* (first published 1957), Barfield Press.

Barrow, J. (2001) *The Book of Nothing,* Vintage.

Bell, J.S. (1987) 'The Challenge of Einstein-Podolsky-Rosen and the Two Voices of Bohr's Response,' from Beller, M. *Quantum Dialogue: the Making of a Revolution.*

Ben Jacob, E., Becker, I., Shapira, Y., Levine, H. (2004) 'Bacterial linguistic communication and social intelligence,' *Trends in Microbiology* Vol. 12 August.

Ben Jacob, E., Shapira, Y., & Tauber A. (2006) 'Seeking the foundation of cognition in bacteria, Schrödinger's negative entropy to latent information', Physica A 359:495–524.

Bockemühl, J. (1981) *In Partnership with Nature,* Bio-dynamic Literature.

Bohm, D. (1980) *Wholeness and the Implicate Order,* Routledge & Kegan Paul.

—, & Hiley, B.J. (1993) *The Undivided Universe: An ontological interpretation of quantum theory,* Routledge.

Bohr, N. (1918) Letter to O.W. Richardson, 15 August, CW, Vol. 3, p. 14.

—, (1919) Letter to A. Sommerfeld, 27 July, CW, Vol. 3, p. 17.

—, (1927) 'The Quantum Postulate and the Recent Development of Atomic Theory,' in *The Philosophical Writings of Niels Bohr Volume 1 Atomic Theory and the Description of Nature* (1987), Ox Bow Press.

—, (1929) Introductory Survey from *The Philosophical Writings of Niels Bohr,* Vol. 1, *Atomic Theory and the Description of Nature* (1987), Ox Bow Press.

—,(1935) 'Quantum Mechanics and Physical Reality,' *Nature* 136:1025–1026.

—, (1938) 'Natural Philosophy and Human Cultures,' in *Nature* 143, 269 from *The Philosophical Writings of Niels Bohr,* Vol. 2, Essays 1933–1957 on Atomic Physics and Human Knowledge (1987), Ox Bow Press.

—, (1956) Draft of letter from Bohr to Heisenberg, never sent in reply to letter of Heisenberg to Robert Jungk, quoted in Niels Bohr Archive, www.nbi.dk/NBA/papers/docs/cover.html.

—, (1958) Discussion with Einstein, from *The Philosophical Writings of Niels Bohr,* Vol. 2, Essays 1933–1957 on Atomic Physics and Human Knowledge, Ox Bow Press.

Boole, G. (1854) *An Investigation into the Laws of Thought on Which are Founded the Mathematical Theories of Logic and Probabilities,* Cambridge University Press.

Born, M. (1926) 'The Structure of the Atom' from Hawking, S. (2010) *Dreams that Stuff is Made of,* Running Press.

Bortoft, H. (1970) 'The Ambiguity of "One" and "Two" in the Description of Young's Experiment,' *Systematics,* December.

—, (1996) *The Wholeness of Nature,* Floris Books.

—, (2010) 'The Transformative Potential of Paradox,' *Holistic Science Journal,* Vol. 1, Issue 1.

—, (2012) *Taking Appearance Seriously: The Dynamic Way of Seeing in Goethe and European Thought,* Floris Books.

Capra, F. (1982) interviewed by Renée Weber in Ken Wilbur, *The Holographic Paradigm,* Shambhala.

—, (1989) 'Howling with the Wolves, Werner Heisenberg,' in *Uncommon Wisdom: Conversations with Remarkable People,* Bantam Books.

Chown, M. (1999) *The Magic Furnace,* Jonathan Cape.

Clarke, C. (2011) 'What Consciousness Does: A Quantum Cosmology of Mind,' *Journal of Cosmology,* Vol. 14.

Cramer, J. (1986) 'The Transactional Interpretation of Quantum Mechanics,' *Reviews of Modern Physics* 58, July (pages 647–88) (online at http://mist. npl.washington.edu/npl/int_rep/tiqm/TI_toc.html).

Dalai Lama, HH the (2005) *The Universe in a Single Atom: The Convergence of Science and Spirituality,* Broadway.

Datta G. & Singh A. (1985) *History of Hindu Mathematics,* Asia Publishing.

Dirac, P. (1931) 'Quantised singularities in the electromagnetic field,' *Proceedings of the Royal Society of London A.* 133, pp. 60–72.

Eckermann, J.P. (1970) *Conversations with Goethe,* John Oxenford, Everyman's Library.

Eddington, A. (1920) *Space Time and Gravitation*, Cambridge Science Classics

—, (1923) *Mathematical Theory of Relativity,* Cambridge University Press.

—, (1935) Meeting of the Royal Astronomical Society, January 11, *The Observatory* 58, pp. 33–41.

Einstein, A. (1905) 'On the electrodynamics of moving bodies,' in *The Principle of Relativity* (1952), Dover.

—, (1916) 'The Foundations of the General Theory of Relativity,' in *The Principle of Relativity (*1952), Dover.

—, (1928) Letter to Schrödinger, on May 31, quoted in Arthur Fine *The Shaky Game: Einstein Realism and the Quantum Theory,* The University of Chicago Press, 1996.

—, (1936) 'Physics and Reality,' in *Ideas and Opinions* (1954), Crown Trade Paperbacks.

—, (1952) *Relativity,* Methuen & Co., Limited.

Einstein, Podolsky, B., Rosen, N. (1935) 'Can Quantum-Mechanical Description of Physical Reality be Considered Complete?' *Physical Review* 47 10: 777–80, 15 May.

Engel G., Calhoun T., Fleming R. *et al.* (2007) 'Evidence for Wavelike energy transfer through quantum coherence in photosynthesis,' *Nature,* 12 April 782–86.

Feynman, R.P., Leighton, R.B., Sands, M., (1965a) *The Feynman Lectures on Physics*, Vol. 3, Addison-Wesley.

—, (1965b) 'The Development of the Space-Time View of Quantum Electrodynamics,' Nobel Lecture, December 11, (http://www.nobelprize. org/nobel_prizes/physics/laureates/1965/feynman-lecture.html retrieved on May 9, 2013).

Fibonacci, L., (1202), *Liber Abaci*, (http://www-history.mcs.st-andrews. ac.uk/Biographies/Fibonacci.html retrieved on July 30, 2015)

Fine, A. (1996) Shaky Game, Einstein Realism and the Quantum Theory, 2nd edition, University of Chicago Press.

Freud, S. (1910) Leonardo da Vinci and a Theory of his Childhood from the Uncanny, (2003) Penguin Books.

Franses, P. (1980) *Something Else, a Journey through Africa*, (personally distributed).

—, (1982) *The Diary of a Computer Programmer*, (personally distributed)

—, (2006) Living Ambiguity, Contextual Choice in the Outcome of DNA, MSc thesis Schumacher College.

—, (2013) 'The Language of Living Processes,' in Lambert *et al.*, *The Intuitive Way of Knowing: A Tribute to Brian Goodwin*, Floris Books.

Gambini R., García-Pintos L., Pullin, J. (2011) 'An axiomatic formulation of the Montevideo interpretation of quantum mechanics,' arXiv:1002.4209v2 [quant-ph].

Gödel, K. (1931) 'On formally undecidable propositions of Principia Mathematica and related systems,' in Kurt Gödel *Collected works*, Vol. I (1986) Oxford University Press: 144–95.

Goethe, J.W. von (1790) *The Metamorphosis of Plants*, The M.I.T. Press (2009).

Goodwin, B. (2007) *Nature's Due: Healing our Fragmented Culture*, Floris Books.

Grandy, D. (2009) *The Speed of Light*, Indiana University Press.

Greenstein, G. & Zajonc, A. (2006) *The Quantum Challenge: Modern Research on the Foundations of Quantum Mechanics*, Jones and Bartlett.

Hadamard, J. (1954) *The Psychology of Invention in the Mathematical Field*, Dover.

Hawking, S. (1988) *A Brief History of Time; from the Big Bang to Black Holes*, Guild Publishing.

Haisch, B. (1999) 'Brilliant Disguise: Light, Matter and the Zero Point Field,' *Science and Spirit* 10: 3.

Hamilton, W. (1844) Letter published in *The London, Edinburgh, and Dublin Philosophical Magazine and Journal of Science*, Vol. xxv, pp. 489–95.

Hartle, J. (2005) 'Excess Baggage,' arXiv:gr-qc/0508001v1, 30 Jul, retrieved September 2014.

Heisenberg, W. (1949) *The Physical Principles of the Quantum Theory*, Carl Eckart and Frank C. Hoyt, Dover.

—, (1956), letter to Robert Jungk, in Robert Jungk *Brighter Than a Thousand Suns – A Personal History of the Atomic Scientists*, Harcourt, pp. 102–04.

Hiley, B.J., (2004) 'Information, quantum theory and the brain' in G. Vitiello (Ed.) *Brain and Being*, Vol. 58, pp. 197–214, John Benjamins Publishing Company.

—, (2011) 'Process, Distinction, Groupoids and Clifford Algebra: an Alternative View of the Quantum Formalism,' in *New Structures for Physics,* ed. Coeke B., Vol. 833, p. 705–50, Springer (http://www.bbk.ac.uk/tpru/BasilHiley/Distinction.pdf accessed 7th April 2015.)

—, (2013) 'The Arithmetic of Wholeness,' *Holistic Science Journal,* Vol. 2, Issue 2.

Hossenfleder, S. (2008) 'A Bi-Metric Theory with Exchange Symmetry,' arXiv: 0807. 2838vi.

Husserl, E. (1907) *The Idea of Phenomenology,* Martinus Nijhoff (1964).

Ifrah, G. (1985) *The Book of Numbers: From One to Zero,* Viking.

Jungk, R. (1958) *Brighter than a Thousand Suns: a Personal History of the Atomic Scientists,* Harcourt Brace.

Kauffman, K.L (1982) 'Sign and Space, Religious Experience and Scientific Paradigms,' Proceedings of the IASWR conference, Institute Advanced Studies, Stony Brook, New York, pp. 118–64.

—, (2015) *Laws of Form – An Exploration in Mathematics and Foundations,* (http://homepages.math.uic.edu/~kauffman/Laws.pdf retrieved December 17 2014.)

Kierkegaard, S. (2004) *The Sickness unto Death,* tr. A. Hannay, in *Sygdommen till Doden* (1849), Penguin Books.

Klein, J. (1992) *Greek Mathematical Thought and the Origin of Algebra,* Dover.

Leibniz, G. (1686) *A Discourse on Metaphysics from Philosophical Writings,* Everyman (1973).

—, (1714) *Philosophical Writings,* Everyman (1973).

Leibniz, G. & Clarke, S. (1717) A Collection of Papers, Which passed between the late Learned Mr. Leibnitz, and Dr. Clarke, In the Years 1715 and 1716, London.

Maxwell, J.C. (1887) *Matter and Motion,* Dover, (1991).

McClure, Ms. (2015) 'Praise Poetry,' (http://msmcclure.com/?page_id=9329 accessed on 24 June 2015.)

Miller, A. (2005) *Empire of the Stars: Obsession, Friendship, and Betrayal in the Quest for Black Holes,* Houghton Mifflin.

Misner, C. Thorne, K. Wheeler, J. (1973) *Gravitation,* W.H. Freeman & Co.

Meier, C.A. (2001) *Atom and Archetype,* The Pauli/Jung letters 1932–1958, Princeton University Press.

Monk, R. (2012) *The Life of J. Robert Oppenheimer,* Jonathan Cape.

Mueller, B. (tr. 1989). *Goethe's Botanical Writings.* Ox Bow Press.

Newton, I. (1686) *Principia Mathematica,* Dover.

—, (1714) 'Right to the Invention of the Method of Fluxions, by some called the Differential Method,' No. 342, p. 173, *The Philosophical Transactions of the Royal Society,* London.

Oppenheimer, J. R. (1965) on the Trinity test. Atomic Archive: http://www.atomicarchive.com/Movies/Movie8.shtml, retrieved May 23, 2008.

Pauli, W. (1948) 'Modern Examples of "Background Physics",' in Meier, C.A., *Atom and Archetype,* Princeton University Press, (2001).

Penrose, R. (2010) *Cycles of Time: an extraordinary new view of the universe,* Bodley Head.

Pirsig, R. (1974) *Zen and the Art of Motorcycle Maintenance,* Bantam Books.

Price, H. (1996) *Time's Arrow and Archimedes' Point,* Oxford University Press.

Rabi, I.I. *et al.* (1969) *Oppenheimer,* Charles Scribner and Sons.

Razjivin, Leupold D. & Novodereshkin V. (1998) in *Photosynthesis: Mechanism and Effects,* Vol. 1, Kluwer Academic Publishers.

Reichenbach, H. (1927) *The Philosophy of Space and Time,* Dover.

Ripalda, J.M. (2009) 'Time reversal and negative energies in general relativity,' arXiv:gr-qc/9906012.

Ross, G.M. (1984) *Leibniz,* Oxford.

Russell, B. (1956) *Portraits from Memory and Other Essays,* Simon and Schuster.

Sartre, J-P. (1965) *Nausea,* Penguin Books.

Sharle, D. (1986) *All Over the Globe: The Past Present and Future of Communication Cables,* tr. by Boria Kuznetsov, Mir Publishers.

Smolin, L. (1988) 'Space and time in the quantum universe' in Ashtekar, A., Stachel, J., *Conceptual Problems of Quantum Gravity,* Birkhäuser.

Spencer Brown, G. (1974) *Only Two Can Play This Game,* Bantam Books.

—, (2009) *Laws of Form,* revised 5th edition, Bohmeier Verlag.

Steiner, R. (1897), *Goethe's World View: The Contemplation of the World of Colours,* (http://wn.rsarchive.org/Books/GA006/English/MP1985/GA006_c03.html.)

Theissen, G. & Saedler H. (2001) 'Plant Biology: Four Quartets,' *Nature* 409: 469–71.

Tsoknyi Rinpoche (2013) *The Best Buddhist Writing,* Shambhala.

Turner, V. (1969) *The Ritual Process: Structure and Anti-Structure,* Aldine de Gruyter.

Weizsäcker, C. von (1980) *The Unity of Nature,* Farrar, Straus & Giroux.

Weyl, H. (1931) *Levels of Infinity,* tr. Peter Pesic, Dover (2012).

—, (1946), 'Mathematics and Logic: A brief survey serving as a preface to a review of *The Philosophy of Bertrand Russell,*' *American Mathematical Monthly* 53: 2–13.

Wheeler, J.A. & Feynman, R.P. (1945) 'Interaction with the Absorber as the Mechanism of Radiation,' *Reviews of Modern Physics* 17 (2–3): 157–61.

Wheeler, J.A. (1998) *Geons, Black Holes and Quantum Foam: A Life in Physics,* W.W. Norton.

Whitehead, A. & Russell, B. (1927), *Principia Mathematica,* Cambridge University Press.

Zajonc, A. (1993) *Catching the Light: The Entwined History of Light and Mind,* Bantam Press.

Index

advanced wave 11, 88f, 192 *see also* retarded wave
Africa 57-62, 126, 224
African 100, 192
aion 208, 210, 218, 220
Alps 101, 126, 224
ambiguity 54, 76f, 86, 104, 141, 203, 206, 211
archetype 36f, 142, 205
Aristotle 108, 130, 133, 205
Aston, Francis 216
atom 20, 28, 38, 45, 56, 120, 122, 186, 196, 205, 209, 216f, 220, 224f

bacteria 133, 147, 203
Barfield, Owen 33, 106, 121, 176
Bell, John S. 85, 116
Big Bang 211, 214
bi-metric 98, 222
— theory of relativity 11
binding energy 216
black hole 119, 124, 145
Bohm, David 9, 43, 94
Bohr, Niels 9, 20, 22, 28, 29, 38, 97, 110, 112, 116, 120, 175, 188, 196
Boole, George 165
Born, Max 29
Bortoft, Henri 9, 43, 45, 55, 68, 92, 138
boundary 31, 115, 146
Brahmagupta 107
Buber, Martin 65

calculus 46, 49, 56

Cantor, Georg 168
carbon 214
cell 133, 152, 154
certainty 52, 57, 79, 170, 171
chance 106, 115, 117, 120f, 124, 191
Chandrasekhar, Subrahmanyan 117, 124
chestnut tree 143, 218, 221
chronological 15, 209
chronos 208f, 220
Clarke, Chris 212
colour 24, 114, 146, 180
complementarity 38, 109, 175
concepts 21, 29, 33, 37, 91, 109, 112, 118, 122–24, 133–36, 157, 175, 185f, 199
Copenhagen 10, 29, 38, 196, 220
cosmological constant 222
Cramer, Jim 93
creation 125, 195, 213f, 218

Dalai Lama 113
dark
—energy 98, 222
— matter 11, 98, 222
darkness 37, 39, 45f, 90, 114, 124, 180, 226
Darwin, Charles 82, 121, 165, 167
decimal system 107, 111
Descartes, René 135
destruction 39, 45, 120, 125, 182, 196, 216, 224
dice 9, 191
dimension 36, 54, 190
distinction 172

Domninus of Larissa 108, 133
double-slit experiment 25, 219
dynamic 31, 115, 133, 137, 218

East 106, 111f
Eddington, Arthur 117, 124, 145, 213
Einstein, Albert 9, 31, 36, 63, 72, 79, 86, 88, 116f, 175, 185
electricity 76, 82
electromagnetism 76, 78
electrons 22, 225
emptiness 107, 111, 130, 160, 178, 201, 222
energy 150, 196, 204, 210, 214, 216
Enlightenment, The 21, 28
eternity 219
event moment 192
Existential Completion Cycle 186
experience 44, 49, 54, 58, 69, 106, 114, 125, 130, 144, 176, 184, 189, 206, 224, 226

Feynman, Richard 11, 26, 89
Fibonacci, Leonardo 107
field 73, 79, 83
final participation 33
Fine, Arthur 175
finite 55, 66, 146, 170, 179, 184
finite boundary 160
finiteness 124
fission 40, 216
force 48, 56, 78, 225
— strong 225
— weak 225
form 58, 94, 100, 111, 133, 138, 148, 156, 160, 173, 193, 204, 218, 221
freedom 54, 59, 76, 100, 124, 138, 143, 184, 225
Freud, Sigmund 80

galaxies 212
Gambini, Rodolfo 198, 202
genes 139, 147, 154
God 47, 54, 163, 181, 219

Gödel, Kurt 165, 170, 186
Goethe, Johann W. von 13, 114, 133, 136, 143, 145, 180, 226
gravity 72, 119, 212

Hamilton-Jacobi 210
Hamilton, William R. 190, 210
Hartle, James 199, 201
Hawking, Stephen 213
Heisenberg, Werner 9, 38, 112, 120, 196
Hertz, Heinrich 85
Hiley, Basil 13, 43, 94
Hoyle, Fred 214
(the) Hub and the rim 160, 209

Incompleteness Theorem 165, 170f
indefiniteness 202
infinite 54f, 168, 178, 187, 204
interference 26, 44

Jacob, Ben 147, 148
journey 7, 57, 69, 127, 140, 154, 220
Judaism 33
Jung, Carl 9, 36, 205, 207

Kauffman, Louis 10, 160, 190 192
key 8, 57, 79
Kierkegaard, Søren 55, 178

language 30, 187
Large Hadron Collider 222
Laws of Form 173
Laws of Thought 165, 167, 173
leaf sequence 139
Leibniz, Gottfried W. von 9, 46, 48f, 51, 54, 56, 64, 71, 73, 187
light 23f, 63, 76, 82
Light in Time solution 76, 139
logic 69
Luxor, temple of 132

magnetism 76, 82
mass 21, 63, 216

mathematics 16, 36, 46, 87, 97, 112, 118,
  134, 161, 165, 168f, 170, 172, 205
Maxwell, James C. 9, 76, 78, 82, 87
meaning 8, 56, 69, 76, 88, 91, 100, 104,
  120, 145, 162, 173, 187, 191f, 197,
  203, 209, 213, 222
measurement 7, 24, 28, 80, 97, 115,
  122, 135, 189, 199
metamorphosis of plants 137
momentum 29, 71, 79, 98, 109, 176, 180
Montevideo Interpretation of
  Quantum Theory 198, 202
motion 21, 48, 78
mythological 15, 70, 194, 209, 218

neutrons 22, 217
Newton, Isaac 9, 46, 50f, 54, 56, 73,
  78, 114, 146, 187
nothingness 24, 37, 39, 106
nucleus 22, 38, 214, 217, 223

observation 27, 66, 135, 145, 188, 212
olam 218
one 108, 130, 133, 161
Oppenheimer, Julius R. 34, 217

paradox 8
part 50
particle 13, 25f, 44, 79, 95, 109, 164,
  172, 180, 210
parts 15, 44, 137, 144, 154
Pauli, Wolfgang 9, 36, 41, 109, 123,
  205, 207
Penrose, Roger 211
phenomenology 10, 55, 180
photon 24f, 44, 98, 125, 145
physics 15, 29, 36, 46, 86, 118, 123f,
  171, 175, 188, 192, 194, 205, 210, 226
Pirsig, Robert M. 81, 130, 163
Podolsky, Boris 116
Poincaré, Henri 170, 184, 186
position 29, 71, 79, 98, 109, 176, 180
potential 56, 86, 96, 133, 160, 176, 203f
pre-impression 83
Pythagoras 134

quality 81, 130, 163, 218
quantum
— potential field 11, 94
— theory 20, 36, 38, 80, 93, 109, 115,
  120, 122, 135, 148, 175, 196, 198,
  201, 210, 225
— time 201, 204
Quaternions 190

relativity 64, 72, 98, 119, 145, 192
retarded wave 11, 88, 89, 192 see also
  advanced wave
Ripalda, Jose M. 98
Rosen, Nathan 116
Royal Society 46, 52f
Russell, Bertrand 165, 169, 171, 173

Sartre, Jean–Paul 165, 182
Schrödinger, Erwin 29
science 15, 34, 38, 47, 77, 82, 119, 121,
  164, 194, 221
sets 168
shore of existence 21, 24, 124
singularity 125, 211
slime mould 133, 156, 204
slits 44, 79, 95
Smolin, Lee 200
space 21, 63, 68, 72–74, 107, 124, 149,
  161, 172
speed of light 9, 63, 66, 79, 84
Spencer Brown, G 10, 160f, 171, 173,
  176, 178, 194
spirit 15, 50, 111, 123
sporulation 149
star 67, 86, 90f, 119, 125, 212, 214
stewardship 158
structure 10, 54, 69, 118, 122, 125,
  175, 187
sūnya 107
symmetry 65, 217

Tagore, Rabindranath 112
Testament, Old and New 218
theology 15, 34, 218
Thompson, George 25

Thompson, Joseph J. 25
time 63, 72, 74, 76, 79, 92, 113, 124, 145, 149, 196–98, 202, 208, 211, 219, 222
Time in Light 76
Transactional Interpretation 93
transition 24
travelling 129
Turner, Victor 70

uncertainty 38, 175

Vieta, François 134
Vishnu the Preserver 217, 219

wave 11, 25, 44, 84, 109, 164, 172, 180
wave function 29, 93
Weizsäcker, Carl Friedrich von 10, 46, 196f, 204, 208, 216

Weyl, Hermann 168f
Wheeler, John 11, 89, 118
wheel of physics 175f, 191
white hole 120, 125
— singularity 213
whole 15, 44, 50, 137, 144, 149, 151, 154, 159, 178, 192, 208
wholeness 7, 43, 45, 50, 69, 97, 133, 208
Wilkinson Microwave Anisotropy Probe (WMAP) 212
world
— clock 36, 45
— harmony 36, 41

Zajonc, Arthur 24, 26, 110
zero 106, 111f, 118, 161

Holonomics
Business where People and Planet Matter
Simon Robinson and Maria Moraes Robinson

Businesses around the world are facing rapidly changing economic and social situations. Business leaders and managers must be ready to respond and adapt in new, innovative ways.

The authors of this groundbreaking book argue that people in business must adopt a 'holonomic' way of thinking, a dynamic and authentic understanding of the relationships within a business system, and an appreciation of the whole. Using real-world case studies and practical exercises, the authors guide the reader in a new, holistic approach to business, towards a more sustainable future where both people and planet matter.

Simon Robinson is a consultant in innovation, strategy and complexity. He has a Masters degree in Holistic Science from Schumacher College, UK.

Also available as an eBook

florisbooks.co.uk

Grow Small, Think Beautiful
Ideas for a Sustainable World
from Schumacher College

Edited by Stephen Harding

Schumacher College, based near Totnes in Devon, England, opened its doors in the early 1990s and is now an internationally-renowned centre for transformative learning on all aspects of sustainable living.

James Lovelock led the first course on Gaia theory. A host of visionary thinkers has followed, including mathematician and biologist Brian Goodwin, who died in 2009. This book is a realisation of his vision for Schumacher College to publish a collection of essays on sustainable solutions to the current global crisis. Themes include the importance of education, science, Transition thinking, economics, energy sources, business and design, in the context of philosophy, spirituality and mythology.

The contributors include Satish Kumar, Jules Cashford, Fritjof Capra, Rupert Sheldrake, James Lovelock, Peter Reason, Gideon Kossoff, Craig Holdrege, Helena Norberg-Hodge, Colin Tudge, Nigel Topping and many others.

 Also available as an eBook

florisbooks.co.uk

First Steps to Seeing

A Path Towards Living Attentively

Emma Kidd

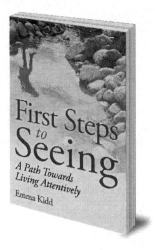

In the twenty-first century we are confronted with a rapidly changing world full of social, economic and environmental uncertainties. We are all inherently connected to this changing world and in order to create the best possible conditions for life to thrive, we must each develop an inner capacity to respond and adapt to life in new, creative and innovative ways.

First Steps to Seeing reveals a practical set of stepping stones that guide the reader into this dynamic way of seeing and relating. Using personal stories, practical exercises and real-world case studies in development, education and business, the author takes the reader on a journey to explore how to give our full attention to life, and how to enliven the world that we each co-create. An inspiring guide for all those working for social change in youth work, business, education or research, or simply seeking fresh paths in life.

Emma Kidd is an educator, writer, independent researcher and consultant. She has a Masters degree from Schumacher College, UK, where she specialised in Phenomenology and the work of Henri Bortoft.

 Also available as an eBook

florisbooks.co.uk

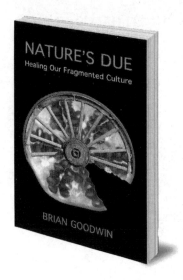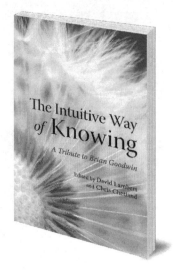

Nature's Due
Healing Our Fragmented Culture
Brian Goodwin

Nature's Due challenges modern ideas on the interaction of science, nature and human culture, with far-reaching consequences for how we govern our world.

The Intuitive Way of Knowing
A Tribute to Brian Goodwin
Edited by David Lambert & Chris Chetland

The Intuitive Way of Knowing is a collection addressing Brian Goodwin's groundbreaking work on evolutionary and theoretical biology.

Professor Brian Goodwin (1931-2009) was born in Montreal and studied biology at McGill University before reading mathematics at the University of Oxford and doing a PhD at the University of Edinburgh with C.H. Waddington. He taught Holistic Science at Schumacher College in Devon, UK.

florisbooks.co.uk